PHARMACOLOGY - RESEARCH, SAFETY TESTING AND REGULATION

PACLITAXEL

BIOSYNTHESIS, MECHANISMS OF ACTION AND CLINICAL USES

PHARMACOLOGY - RESEARCH, SAFETY TESTING AND REGULATION

Additional books in this series can be found on Nova's website under the Series tab.

Additional e-books in this series can be found on Nova's website under the e-book tab.

CANCER ETIOLOGY, DIAGNOSIS AND TREATMENTS

Additional books in this series can be found on Nova's website under the Series tab.

Additional e-books in this series can be found on Nova's website under the e-book tab.

PHARMACOLOGY - RESEARCH, SAFETY TESTING AND REGULATION

PACLITAXEL

BIOSYNTHESIS, MECHANISMS OF ACTION AND CLINICAL USES

DIEGO MORALES
EDITOR

NOVA BIOMEDICAL

New York

For permission to use material from this book please contact us:
Telephone 631-231-7269; Fax 631-231-8175
Web Site: http://www.novapublishers.com

Library of Congress Cataloging-in-Publication Data

ISBN: 978-1-62808-549-5

LCCN: 2013944378

Published by Nova Science Publishers, Inc. † New York

Contents

Preface

Paclitaxel is an antimitotic drug that promotes microtubule polymerization and produces sensitization to radiation. Paclitaxel is a first-line drug for the treatment of non-small cell lung cancer and is used to treat metastatic ovarian cancer, breast cancer, head and neck cancer, and advanced cases of Kaposi's sarcoma. In this book, the authors discuss the biosynthesis, mechanisms of action and clinical uses of paclitaxel. Topics include paclitaxel resistance and overcoming strategies with structurally or functionally related drugs; molecular mechanisms that mediate resistance to paclitaxel in cancer; the biosynthesis of paclitaxel and its pharmaceutical application for cancer treatment; current status and clinical trials of paclitaxel-containing therapy for urothelial cancer; mesenchymal stromal cells used for carrying and delivering paclitaxel; and the role of paclitaxel in ovarian cancer.

Chapter 1 - Paclitaxel is established in the treatment of malignant tumors as a potent chemotherapy agent. However, acquired multi-drug resistance (MDR) limits its therapeutic usefulness, which is a common problem for the efficient chemotherapy. Over-expression of the *mdr1* gene product P-glycoprotein (P-gp) was extensively studied *in vitro* in association with paclitaxel resistance as well as the alterations in the expression of tubulins or microtubule associated proteins. In addition, the correlations between paclitaxel-resistance and mutated *p53*, over-expressed Bcl-2, Bcl-xL or NFkB as well as down-regulation of HER-2 were also examined. However, these data are difficult to translate in clinic mainly due to study/patient heterogeneity and the fact that no convincing useful biomarkers exist that can predict resistance to paclitaxel in cancer patients. This is an emerging issue underneath the principle of individualized cancer therapy. Despite these problems, taxanes represent one of the most promising classes of anticancer agents. The notable achievements with paclitaxel in clinical oncology stimulated the identification of new derivatives with improved activities in resistant cancer cells, resulting in a huge number of novel natural and synthetic taxanes. Some of them showed a promising behavior in the presence of MDR phenotype retaining also the activity against cells bearing alterations of the tubulin/microtubule system. Microtubule-stabilizing compounds can be classified into three groups: (i) the terpenoids, including paclitaxel, docetaxel, and eleutherobin; (ii) the macrolides, including peloruside, epothilone, and laulimalide; and (iii) the polyhydroxylated alkatetraene lactones, including discodermolide. Among them, particularly promising are agents that do not interact with the P-gp, such as epothilone, discodermolide and laulimalide. Recently, it was shown that macrocyclic diterpenes of jatrophane type interact with tubulin/microtubule system and exhibit considerable anti-P-gp activity against paclitaxel resistant cancer cells. In this chapter, the authors describe the main molecular mechanisms of paclitaxel resistance identified to date, focusing on the development of new microtubule interacting taxanes or other structurally distinct drugs potentially useful for the treatment of resistant tumors.

Chapter 2 - Paclitaxel (Taxol) is an antimitotic drug that promotes microtubule polymerization and produces sensitization to radiation. Paclitaxel is a first-line drug for the treatment of non-small cell lung cancer and is used to treat metastatic ovarian, breast cancer, head and neck cancer, and advanced cases of Kaposi's sarcoma. The development of paclitaxel resistance, which reduces the efficacy of paclitaxel in a number of diseases, has been discussed in several review articles. Paclitaxel resistance can result from an increased efflux via ATP-binding cassette (ABC) multidrug transporters, changes in microtubule dynamics, genetic polymorphisms, alterations in extrinsic and intrinsic apoptotic pathway, activation of transcription factors and altered target or binding sites. However, the pharmacokinetic and cellular resistance mechanisms of paclitaxel are not fully understood. Thus, elucidating paclitaxel resistance mechanisms would significantly facilitate the development of therapeutic interventions to surmount paclitaxel resistance. This chapter discusses the current knowledge about paclitaxel resistance at the molecular level.

Chapter 3 - Pacific yew (*Taxus brevifolia*) is a common tree of in the coniferous forest understory of western North America. All parts of the plant are poisonous except the red layer of fruits (*arilus*). The last thirty years, the pharmaceutical company are interested in this plant due to the presence of the bioactive compounds such as taxine alkaloids (Taxine A and B) and paclitaxel

(Taxol). The chemical name for paclitaxel is 5β,20-Epoxy-1,2α,4,7β,10β,13α-hexahydroxytax-11-en-9-one 4,10-diacetate 2-benzoate 13-ester with (2R,3S)-N-benzoyl-3-phenylisoserine.The concentration of paclitaxel is 0.01-0.02% in the bark of this yew. The structure of paclitaxel is very complex. From this reason and the complicated procedures for the synthesis, the researchers rather isolate this compound from the bark of yew. The pathway of paclitaxel biosynthesis has 20 steps. The main precursor for biosynthesis is geranylgeranyl diphosphate, which is cyclizated to taxa-4(5),11(12)-diene. The obtained olefin is oxidized to taxa-4(20),11(12)-dien-5α-ol by the cytochrome P450, and then the alcohol is converted to the corresponding acetate ester by the acetyl CoA. After confirmation of the anticancer activity of paclitaxel, the World Health Organisation allowed its application in the treatment of cancer. Paclitaxel simulates the assembly of tubulin dimmers in microtubules. It stabilizes the microtubules and prevents their depolymerization. This induced stability inhibits the normal dynamic reorganization of the microtubules network. It blocks cells in the G2/M phase of the cell cycle and such cells are unable to form a normal mitotic apparatus. The clinical studies were shown that the paclitaxel is effective in the cancer therapy of ovarian, intestine, lung and breast. Today, a few commercial drugs with the paclitaxel as a pharmaceutical active compound are available in the market. The preparations are commonly administered intravenously. It is supplied as a non-aqueous solution intended for dilution with a suitable parenteral fluid prior to intravenous infusion. Based on the successful application in the cancer treatment, the paclitaxel has the significant place among the other chemotherapeutic agents even today.

Chapter 4 - Urothelial cancer (UC) arising from the urinary bladder, renal pelvis, or ureter is a common cancer in adult males. In general, UC is recognized as chemo-sensitive and patients with advanced-stage lesions including metastatic UC are treated using chemotherapy-based methods. At present, combination therapy with gemcitabine and cisplatin (CDDP) is standard in patients with such advanced UC. MVAC therapy comprising methotrexate, vinblastine, doxorubicin, and CDDP is also used for these patients. However, unfortunately, most of these patients experience relapse and no standard salvage therapies have been devised for use after CDDP-containing regimens. In recent years, various new chemotherapeutic agents and regimens have been reported as effective and safe for CDDP-resistant UC, and various targeted and anti-angiogenic therapies are also performed. However, most of these approaches have been described in preliminary studies and further investigations are necessary to confirm their clinical feasibility.

Clinical studies are thus in progress all over the world. Paclitaxel (PTX) appears effective against UC. Previous reports have identified PTX monotherapy as an active first line chemotherapeutic regimen. In addition, this agent is considered a good option for salvage therapy, because it is well tolerated and safe. On the other hand, whether PTX monotherapy or PTX-based combination chemotherapy offers satisfactory anti-tumoral effects and prolonged survival in patients with CDDP-resistant UC remains unclear. The aim of this study was to clarify the current status of PTX-containing chemotherapy as second line therapy for patients with UC. In addition, the authors introduce clinical trials of PTX-containing chemotherapies and combination therapies with PTX and molecular-targeted therapy for advanced UC. They also describe delivery systems and intravesical therapies using PTX. This chapter therefore introduces comprehensive and wide-ranging information regarding the current status of PTX-based therapies in patients with UC.

Chapter 5 - In recent years many strategies have been developed to obtain high concentrations of anti-cancer drugs effective on tumor mass. Genetically modified Mesenchymal Stromal Cells (MSCs) have been indicated as a candidate to deliver such drugs.

In a previous study, the authors found that mouse-bone-marrow-derived MSCs (BM-MSCs) exposed to high doses of Doxorubicin were able to incorporate and release the drug, acquiring an anti-proliferative potential towards adjacent co-cultured haematopoietic stem cells. The authors thus hypothesized whether freshly isolated human BM-MSCs, adipose tissue MSCs (AT-MSCs) and skin-derived stromal fibroblasts (hSDFs) primed *in vitro* with Paclitaxel (PTX) may acquire anti-tumor activity when localized near cancer cells.

Incorporation of PTX into MSCs was studied by using FITC-labelled-PTX and analyzed by FACS and confocal microscopy. Release of PTX in culture medium by PTX primed BM-MSCs (BM-MSCsPTX) was investigated by HPLC. The authors showed that BM-MSCs, AT-MSCs and hSDFs acquired strong anti-tumor capacity, once primed with PTX. Cells were able to uptake PTX and release it in a time-dependent manner and the released drug was active *in vitro* against proliferation of leukemia, osteosarcoma, prostatic carcinoma and neuroblastoma cell lines.

In vitro, both BM-MSCs and BM-MSCsPTX bound and formed rosettes with leukemia cells (LCs). Ultrastructural analysis of rosettes did not show morphological alterations of BM-MSCsPTX, while the attached LCs were apoptotic and necrotic. BM-MSCs and BM-MSCsPTX released molecules that

reduced LCs adhesion to microvascular endothelium (HMECs) and down-modulated ICAM-1 and VCAM-1 on HMECs. *In vivo* studies showed that BM-MSCsPTX had a strong anti-tumor activity if co-injected with tumor cells or injected intra-tumor. Established subcutaneous tumor nodules in mice treated with BM-MSCsPTX showed a significant size reduction as well as decreased vascularisation. Priming human MSCs with PTX is a simple procedure that does not require any genetic cell manipulation. The harvesting of BM-MSCs, AT-MSCs and hSDFs do not require invasive procedures and is without any particular pain or discomfort to donor patients. The authors' studies demonstrate for the first time that MSCs isolated from different tissue sources can uptake and subsequently slowly release PTX. This suggests that MSCs could become a tool for drug delivery due to its ability to increase efficacy and target selectivity in cancer therapy.

Chapter 6 - Epithelial ovarian carcinoma, although rare, causes more deaths than any other female genital tract cancer and is overall the fifth most frequent cause of cancer death in women. Its high lethality is because more than two-thirds of patients have a highly advanced disease at diagnosis. In the United States, there were approximately 22,280 new cases and 15,500 deaths in 2012 as a result of ovarian cancer. Systemic chemotherapy plays an important role in the treatment of epithelial ovarian cancer. After the introduction of cisplatin in the latter half of the 1970s, platinum-based combination chemotherapy became the regimen of choice. The incorporatior of paclitaxel into chemotherapeutic regimens in late 1990s was a majc advance in the treatment of advanced stage disease. A series of randomize prospective clinical trials with paclitaxel-containing arms defined t combination of intravenous carboplatin and paclitaxel as the standard for fi line chemotherapy for ovarian cancer, to the point where now ovarian car treatment cannot be started without paclitaxel.

In: Paclitaxel
Editor: Diego Morales
ISBN: 978-1-62808-549-5

Chapter 1

Paclitaxel Resistance and Overcoming Strategies with Structurally-or Functionally-Related Drugs

Jasna Banković, Ana Podolski-Renić and Milica Pešić*
Institute for Biological Research "Siniša Stanković",
University of Belgrade, Belgrade, Serbia

Abstract

Paclitaxel is established in the treatment of malignant tumors as a potent chemotherapy agent. However, acquired multi-drug resistance (MDR) limits its therapeutic usefulness, which is a common problem for the efficient chemotherapy. Over-expression of the *mdr1* gene product P-glycoprotein (P-gp) was extensively studied *in vitro* in association with paclitaxel resistance as well as the alterations in the expression of tubulins or microtubule associated proteins. In addition, the correlations [illegible] paclitaxel-resistance and mutated *p53*, over-expressed Bcl-2, or NFkB as well as down-regulation of HER-2 were also examined. However, these data are difficult to translate in clinic mainly

* Corresponding author: Dr. Milica Pešić. E-mail: vtbvtm@sbb.rs.

due to study/patient heterogeneity and the fact that no convincing useful biomarkers exist that can predict resistance to paclitaxel in cancer patients. This is an emerging issue underneath the principle of individualized cancer therapy. Despite these problems, taxanes represent one of the most promising classes of anticancer agents. The notable achievements with paclitaxel in clinical oncology stimulated the identification of new derivatives with improved activities in resistant cancer cells, resulting in a huge number of novel natural and synthetic taxanes. Some of them showed a promising behavior in the presence of MDR phenotype retaining also the activity against cells bearing alterations of the tubulin/microtubule system. Microtubule-stabilizing compounds can be classified into three groups: (i) the terpenoids, including paclitaxel, docetaxel, and eleutherobin; (ii) the macrolides, including peloruside, epothilone, and laulimalide; and (iii) the polyhydroxylated alkatetraene lactones, including discodermolide. Among them, particularly promising are agents that do not interact with the P-gp, such as epothilone, discodermolide and laulimalide. Recently, it was shown that macrocyclic diterpenes of jatrophane type interact with tubulin/microtubule system and exhibit considerable anti-P-gp activity against paclitaxel resistant cancer cells. In this chapter, we describe the main molecular mechanisms of paclitaxel resistance identified to date, focusing on the development of new microtubule interacting taxanes or other structurally distinct drugs potentially useful for the treatment of resistant tumors.

Introduction

Paclitaxel is one of the most promising anti-cancer agents, which acts as a microtubule-stabilizing agent. Since its mechanism was figured out in the late 1970s, paclitaxel has drawn the attention of both physicians and researchers. Paclitaxel acts as a microtubule-stabilizing agent. It binds reversibly to β-tubulin, which stabilizes microtubule complexes and promotes microtubule polymerization [1]. The prevention of microtubule disassembly and the suppression of their dynamics disrupt the mitotic spindle, arresting the cell cycle at the metaphase-anaphase transition and leading to apoptosis [2, 3]. Although the primary mechanism of paclitaxel is clear, the changes in signaling cascade promoted by paclitaxel continue to be explo

Taxanes, paclitaxel and closely related drug – docetaxel a
in clinics for the treatment of several types of tumors, i
breast, head and neck, lung, and prostate cancer [5]. Pacli

from the bark of the Pacific yew, *Taxus brevifolia*, and identified as a potent cytotoxic agent in a screen of natural products sponsored by the National Cancer Institute [6]. In an effort to discover compounds with similar activity as paclitaxel, French scientists synthesized docetaxel from a naturally occurring 10-deacetylbaccatin III, taken from the needles of European yew, *Taxus baccarat* [6]. Although paclitaxel and docetaxel are among the most effective chemotherapeutics, their clinical application is dealing with serious problems, such as water insolubility, development of resistance and deleterious side effects that include neutropenia, neurotoxicity, cardiac effects, and immuno-suppression [8, 9].

Because of its aqueous insolubility, paclitaxel (Taxol®) is dissolved in Cremaphor EL. This carrier molecule often induces hypersensitivity reactions in patients, which are therefore pre-medicated with corticosteroids and antihistamines [9].

Taxanes are susceptible to several mechanisms of drug resistance. Resistance to taxanes may occur prior to exposure (*de novo*, or inherent resistance), as well as during the treatment (acquired resistance). In addition, the development of cross-resistance or multi-drug resistance (MDR) after exposure to structurally and functionally unrelated drug usually results in the resistance to taxanes [10].

The MDR phenotype is often mediated by the over-expression of drug efflux pumps. P-glycoprotein (P-gp) is the best-studied membrane transporter belonging to ATP-Binging Cassette (ABC) transporters. It extrudes the drugs at the membrane and prevents their accumulation within resistant cells [11, 12].

Although MDR is the most extensively reported mechanism of resistance to taxanes, recent studies revealed that changes of the microtubule structure or composition could lead to a reduced sensitivity to taxanes [13]. Thus, taxane resistance can also develop from β-tubulin gene mutations, over-expression of βIII-tubulin or microtubule-associated proteins, and alterations in mitotic checkpoint signaling proteins [14, 15].

Several less studied, but recognized mechanisms of resistance have also been described. They include alterations in signaling pathways, altered regulation of the cell cycle and altered control of apoptosis and cell death signals [16].

The important contribution of taxanes in improving the quality of life and overall survival of cancer patients has stimulated intensive efforts to elucidate the molecular mechanisms of resistance to taxanes and to develop novel agents effective against resistant tumors with fewer side effects [17]. A number of novel agents targeting microtubules have been recently developed.

It is intriguing that almost all the tubulin-interactive agents are natural products or analogues of natural products. They are complex compounds with significant stereochemistry. Made by nature, they are well adapted for binding to the three-dimensional surface of a protein, such as tubulin probably, as a part of unique defense mechanism present in plants or lower species.

Important question in the development of new drugs is whether their supplies could be assured if they are going to be used clinically on a large scale. In the case of paclitaxel, the major initial supplies came from harvesting yew trees, but this was supplanted semi-synthetically [18]. Paclitaxel is also being produced by plant tissue culture techniques [19]. Fortunately, a total synthesis approach proved to be successful in the case of other tubulin-interactive agents, such as the halichondrin B [20, 21]. Therefore, we believe that a proper protocol will be established in order to produce any natural product or its derivative in sufficient amounts required for clinical use.

For sure, there is a perspective for discovering new natural products with tubulin-interactive properties that are less toxic and more effective against resistant tumors. However, the development of new analogs of paclitaxel, the epothilones, and other compounds such as discodermolide will continue to improve the development of chemotherapy.

Mechanism of Paclitaxel Action

Tubulin is a protein that is the central constituent of the microtubule cytoskeleton and has one of the most fundamental cellular functions – the formation of the mitotic spindle, allowing for the equal division of chromosomes during mitosis [22]. Spindle poisons belong to the oldest and most successful chemotherapeutics, which bind tubulin and cause a prolonged mitotic arrest by interfering with the normal dynamic activity of microtubules. This arrest is widely believed to be central to spindle poison–induced cell death, but the mechanisms by which these drugs destroy cells remain unclear.

During mitosis, cells replicate their genetic material and divide it equally into two daughter cells. Interfering with this process can generate cells with irregular chromosome content (a condition known as aneuploidy), as well as genomic instability, cell death, and possibly cancer [23-26]. Cells have developed numerous mechanisms to ensure that mitosis is accomplished with great fidelity. First research reports revealed that when cells were incubated with spindle poisons the number of cells in mitosis dramatically increased over time. Over the last twenty years, researchers have identified the spindle assembly checkpoint (SAC) as the signaling pathway responsible for this mitotic arrest, which ensures accurate segregation of chromosomes. The signal emerges from kinetochores, which form the protein link between microtubules and DNA on each chromatid [27-29]. The activation of cyclin-dependent kinase 1 (Cdk1) by cyclin B drives a cell into mitosis, at which point a complex called cohesin holds the sister chromatids together in a mitotic chromosome. In a normal mitosis, the SAC is active only briefly (a period of minutes) as unattached chromosomes attach to spindle microtubules and correct improper attachments. During this time, SAC activity maintains high cyclin B levels through inhibition of an E3 ubiquitin ligase that normally targets cyclin B for degradation. This ubiquitin ligase is the anaphase promoting complex or cyclosome (APC/C). When all chromosomes are properly attached, the SAC is turned off and the APC/C degrades cyclin B. This drop in cyclin B levels results in exit from mitosis, and both Cdk1 and a protein called securin (which protects cohesin) are targeted (by the APC/C) for degradation [30]. In contrast, when cells enter mitosis in the presence of spindle poisons, kinetochores are never able to form proper attachments between DNA and spindle microtubules.

This results in the permanent activation of the SAC and a mitotic arrest that lasts for hours. These cells will eventually either undergo apoptosis or exit mitosis by an alternate mechanism that does not occur during a normal mitosis. Thus, the SAC is briefly active during a typical mitosis until all kinetochores are attached to spindle microtubules. The SAC also frequently responds to attachment defects that occur under normal conditions by arresting cells in mitosis until errors are resolved. However, spindle poisons greatly extend the duration of this arrest and prevent cells from exiting mitosis correctly. The ubiquitin ligase activity of the APC/C requires activation by Cdc20 [31]. When the checkpoint is active, two SAC proteins called Mad2 and BubR1 (assisted by Bub1) prevent APC/C activation by binding the Cdc20-APC/C complex. When all kinetochores have proper attachments and the SAC is inactive, Cdc20 is free to activate the APC/C, quickly leading to cyclin B degradation and anaphase onset. Importantly, even during a robust SAC response, cyclin B degradation still takes place slowly owing to incomplete inhibition of the APC/C. This phenomenon, known as “slippage”, means that cells treated with spindle poisons can eventually exit mitosis despite an active SAC [32]. Apoptosis, or programmed cell death, occurs in response to a number of intracellular and extracellular stimuli [33]. Some studies linked the apoptotic network to mitotic duration and suggested that maximizing the time that cells spend in mitosis could be a goal of future therapies [34]. However, until recently a mechanism through which the cell would actually monitor the duration of mitosis and link this to the activation of cell death pathways has not been reported. New studies have identified a pathway involving the protein Mcl1 that couples the timing of mitosis to the induction of apoptosis [35]. Mcl1 is a member of newly discovered apoptotic pathway that is active in mitosis and a unique member of the Bcl-2 family of pro-survival proteins [36, 37]. It binds to pro-apoptotic proteins Bak and Bax and inhibits their association with the mitochondrial membrane, so Bak and Bax cannot form pores in the mitochondrial membrane and initiate apoptosis. The expression of Bcl-2 remains constant and only decreases when cells approach terminal differentiation, while Mcl1 expression appears to be temporarily increased at critical periods, including mitotic arrest during treatment with spindle poisons [38]. This enables the cell to better inhibit the apoptotic pathway in the face of powerful insults, which would otherwise cause the immediate induction of cell death. However, because the Mcl1 elevation is transient, apoptosis is eventually initiated if the cell does not correct the problem [35]. When Mcl1 expression is induced after mitotic arrest, several important factors regulate its degradation [35]. FBW7, the substrate-binding component of an ubiquitin ligase complex, targets Mcl1 for degradation by the 26S proteasome, whereas the deubiquitinase USP9X counteracts the polyubiquitinating activity of the FBW7 complex [39]. During mitotic arrest, the activities of Jun N-terminal kinase (JNK), the mitogen-activated protein kinase (MAPK) family member p38, and casein kinase II (CKII) are increased, leading to phosphorylation of Mcl1, interaction with FBW7, and the subsequent polyubiquitination of Mcl1. Polyubiquitylation of Mcl1 leads to degradation via the proteasome and continually decreasing concentrations of Mcl1 during mitotic arrest. If the arrest is alleviated, then JNK, p38, and CKII activity quickly falls and PP2A rapidly dephosphorylates Mcl1, abrogating its interaction with FBW7 and keeping Mcl1 amounts in the cell high. The peak levels of Mcl1 occur immediately following the addition of spindle poisons (at the onset of mitotic arrest), providing time for the cell to resolve errors in kinetochore-microtubule attachment. During this period, the amount of Mcl1 slowly decreases. Concurrently, although the SAC is active cyclin B amounts still slowly decline because of slippage-mediated degradation by the APC/C. If the SAC can be silenced during this period via the attachment of kinetochores to spindle microtubules, Mcl1 levels will be stabilized and cyclin B will be degraded, leading to mitotic exit. However, if (in the case of prolonged treatment with spindle poisons) kinetochores never correctly attach to spindle microtubules and the SAC remains active, the rate at which Mcl1 and cyclin B are degraded ultimately determines the fate of the cell. If cyclin B levels first fall below the threshold for mitotic exit, then the cell will enter anaphase before cell death is initiated. However, if Mcl1 is degraded past the levels necessary to inhibit Bak and Bax, the apoptotic pathway will be triggered before the cell can exit mitosis [40]

The initial discovery of paclitaxel was accompanied with problems, because of compound supply and solubility, and because it showed only relatively modest *in vivo* activity against the current antileukemic models at the National Cancer Institute – NCI in the USA [41]. Fortunately, the B16 melanoma solid tumor assay was introduced in the early 1970s, and paclitaxel showed excellent and reproducible activity against this solid tumor, with an increase in life span (ILS) of 126%, 32%, and 86% in three separate experiments [41].

In 1978, *in vivo* studies of paclitaxel were carried out, and these were rewarded when it was found to show strong activity against the MX-1 breast xenograft; this discovery helped in maintaining interest in the compound [41]. The final significant finding came with the discovery by Horwitz that paclitaxel promoted the assembly of microtubules [42]. This new mechanism was a crucial factor in raising interest in this compound. The solubility problem was overcome by the development of an emulsion formulation in Cremophor, a polyethoxylated castor oil, which unfortunately caused allergic reactions in some patients. To overcome these problems premedication with antihistamines was used and the intravenous infusions, which initially lasted for 24 h, was reduced to 3 h [43].

Paclitaxel, previously known as taxol, was found to have clinical activity against ovarian cancer in 1989 [44] and against breast cancer in 1991 [45]. Bristol- Myers Squibb (BMS) took over its further development in 1991 under a Cooperative Research and Development Agreement (CRADA) with the NCI. BMS was able to trademark the name Taxol for their formulation of the drug, and the generic name paclitaxel was applied to the chemical compound formerly known as taxol. Taxol was approved by the FDA in December 1992 for the treatment of refractory breast cancer and refractory ovarian cancer and was launched on the U.S. market by Bristol-Myers Squibb the following year. Later, it was approved for treatment of breast cancer after failure of combination chemotherapy for metastatic disease, for second-line treatment of AIDS-related Kaposi‘s sarcoma, and for NSCLC in combination with cisplatin [46].

Paclitaxel binds the mitotic spindle microtubules and prevents their polimerization, impeding normal progress of cell division [47]. Microtubules are hollow, cylindrical tubes which consist of α and β heterodimers (Figure 1). These heterodimers connect to form linear protofilamente and 13 laterally associated protofilament microtubules. Microtubules have important role in numerous cellular activities such as maintenance of cell shape, motility, intracellular transport and formation of mitotic spindle during cell division [5]. These are highly dynamic structures that continuously polymerize, i.e. α and β heterodimers incorporate into microtubules and depolymerise, that is, α and β heterodimers are released into the cytoplasm [48] (Figure 1). Dinamic instability is regulated by the numerous proteins associated with microtubules (MAPs) which bind and stabilize the microtubules. MAPs consist of several subtypes: MAP1A, MAP1B, MAP2, MAP4 and tau. Phosphorylation of MAPs is important postranslational modification which influences their separation, leading to microtubules' destabilization [49]. Oncoprotein 18/statmin is another protein which influences the microtubules' stability. Statmin regulates mitotic spindle microtubules by binding to their ends and increasing their depolimerization [50].

Target molecule for paclitaxel is β-tubulin (Figure 1), whereby taxan-binding place exists only on polimerized microtubules. Paclitaxel achieves its cytotoxic effect by binding to mitotic spindle microtubules and preventing their depolimerization, which induces cell cycle arrest in G2/M phase of cell cycle [51] (Figure 1).

This leads to the activation of JNK/SAPK signaling pathway. JNK phosphorylates and inhibits antiapoptotic protein Bcl-2, which initiates apoptosis [52].

During interphase, paclitaxel can also cause disruption of microtubules thereby disrupting growth and metabolism. Some of the most investigated and well-characterized mechanisms of paclitaxel on molecular level are:

1) Activation of cell division control-2 kinase (cdc-2) [53];
2) Stabilization of cyclin B-1 [54];
3) Activation of the spindle assembly checkpoint [55];
4) Induction of apoptosis through phosphorylation of bcl-2 [56];
5) Inhibition of cell proliferation [57].

In triggering apoptosis, both the dose and the duration of exposure appear to be important. At sub-nanomolar concentrations, paclitaxel favors microtubule assembly through reduction of the critical concentration of tubulin dimmers, GTP and microtubule associated proteins (MAPs) [58]. At picomolar concentrations, it also exerts its effects on interphase microtubules and genes controlling apoptosis [59]. Increasing concentrations result in shifting the equilibrium of dimers to polymers thereby preventing disassociation [60].

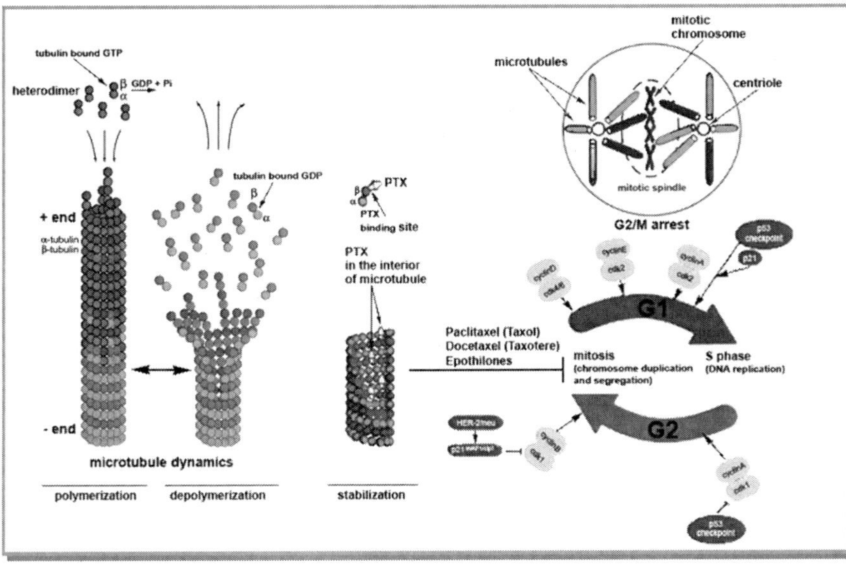

Figure 1. Mechanism of paclitaxel (PTX) action. PTX promotes polymerization of microtubules and their stabilization. It binds to β-tubulin in αβ-tubulin heterodimers and its binding site is localized at the microtubules interior. Besides PTX, other microtubule-stabilizing agents, such as docetaxel and epothilones, induce mitotic arrest in G2/M phase of cell cycle by preventing the segregation of chromosomes. Cell cycle is regulated by complex interaction between cyclins and cyclin-dependent kinases (cdk) as well as by tumor suppressor p53 and downstream signalization from membrane receptors, such as HER2/neu.

However, there are several possible outcomes after cell exposure to paclitaxel. Firstly, the cells are arressted in G2/M phase of cell cycle during paclitaxel presence. This enables cells to survive and continue to devide as diploid (2N) cells [61]. Contrary to this, it could come to mitotic slippage or adaptation, when the cells end the division without anaphase and telophase, whereby tetraploid cells (4N) with no separation of chromosomes arise [62]. These cells can survive and continue to devide as tetraploid (4N) (Adaptation 1). Contrary to this, adapted cells can leave G1 phase of cell cycle, entering into senescence or apoptosis (Adaptation II). Also, cell can avoid control system of G1 phase of cell cycle which leads to apoptosis in interphase (Adaptation III). Finally, apoptosis can be initiated right after cell arrest in G2/M phase of cell cycle [63].

Resistance to Paclitaxel

Paclitaxel has been found to be efficacious in the treatment of a number of human cancers including ovarian cancer, breast cancer, NSCLC, and other malignancies [64-70]. However, it has become obvious that many patients treated with paclitaxel present de-novo or will acquire resistance to this agent.

Numerous factors influence the paclitaxel resistance development including mutations in α and β-tubulins, changes in the composition of β-tubulin isoforms, increased P-glycoprotein expression and changes in the MAPs expression. Also, the disturbance of the normal functioning of some signaling pathways such as cell cycle control, cell growth and apoptosis can have an important role in the development of resistance to paclitaxel [71].

There are 7 different β-tubulin isoforms (class I, II III, IVa, IVb, V, VI), which expression differes from tissue to tissue. βI and βIV-tubulin isoforms are constituvely expressed, while the other isoforms' expression is tissue specific [72]. Some research suggests that microtubules that consist of α/βIII heterodimers are much less stable, with the increased tendency to polimerize comparing to microtubules composed of other β-tubulin isoforms [5, 48]. Besides, there is evidence that the microtubules composed of α/βIII and α/βIV heterodimers are less sensitive to paclitaxel, i.e. the larger quantity of paclitaxel is needed to stabilize microtubules [5, 48]. This research shows that the paclitaxel effect depends on the composition of β-tubulin isoforms, whereby increased βIII and βIV-tubulin expression leads to the development of resistance to paclitaxel.

The majority of tubulin mutations has been identified in βI-tubulin isoform. Mutations in β-tubulin change the microtubule dynamics and/or disturb the binding sites for paclitaxel, leading to the emergence of resistance [72]. Also, after prolonged exposure to paclitaxel, β-tubulin mutations often accompany loss of heterozygosity (LOH) of normal tubulin allele leading to the emergence of a very resistant phenotype [73]. Resistance to paclitaxel is to a lesser extent connected with the mutations in α-tubulin which influence the binding of MAPs to the microtubules [74].

Tau, one of the MAPs subtypes, enables the binding of tubulin heterodimers and stabilization of microtubules [75]. This protein binds the paclitaxel-binding site on the inside of the microtubules. There is evidence about conjunction between elevated tau expression and resistance to paclitaxel: microtubule assembly in the presence of low tau concentrations enables faster binding of paclitaxel when compared with the high tau concentrations [76]. Oncoprotein 18/stathmin is included in the regulation of mitotic spindle by binding to its ends and leading to the microtubules' depolimerization. Hence, stathmin overexpression decreases the polimerization od mitotic spindle microtubules and leads to the decreased binding of paclitaxel influencing its efficacy.

p53 tumor suppressor has a major role in the cell cycle and apoptosis regulation. However, its role in the resistance development is still unclear. It was reported that some tumors with mutated p53 were more sensitive to paclitaxel [77]. Namely, mutated p53 increased the expression of MAP4 which influences the microtubules' polimerization increasing sensitivity to paclitaxel [78]. Also, cells with mutated p53 are able to evade the control system of G1 phase of cell cycle, which introduces the cell much faster into the G2/M phase of cell cycle where paclitaxel exerts its effect [60]. Contrary to this research, there is evidence about connection between the mutated p53 and paclitaxel resistance. Mutated p53 does not influence the expression of pro-apoptotic Bax, which prevents the initiation of apoptosis and can lead to decreased paclitaxel efficacy [79]. Giannakakou et al., used ovarian cancer cell lines which harbor β-tubulin mutations to investigate paclitaxel resistance in association with p53 expression [80]. In two of the sublines an inactivating TP53 mutation was detected, and in the third subline p53 expression was absent. However, analysis of these cells showed that transduction of wild type TP53 did not alter paclitaxel resistance. Paclitaxel can stimulate macrophages to release tumor necrosis factor-α, which in turn induces apoptosis in the absence of p53 [81].

Considerable interest has been placed on deciphering the events associated with taxane related sensitivity and resistance through the study of apoptosis. Early studies indicated that over-expression of Bcl-2 and Bcl-xL contributed to taxane resistance. Additional studies led to the suggestion about a threshold at which specific genes confer resistance [82]. Other studies indicate that the over-expression of pro-apoptotic genes is associated with paclitaxel sensitivity [83] and yet others that show no correlation of Bcl-2 levels and response [84]. As most of the anti-apoptotic genes (IAP, TRAF, Bcl-2, Bcl-xL) are under the transcriptional control of NFkB this has also been studied. In fact, it appears from a few studies that NFkB is constitutively activated in many breast cancers and that inhibition of NFkB may sensitize cells to taxanes [85-87]. In addition, it has been shown that approximately half of breast cancers have increased levels of Akt, which appears to activate Bcl-2 and increase the activation of NFkB. Akt is also activated by HER-2 signaling and is implicated in chemoresistance mechanisms of taxanes [88, 89].

The spindle assembly checkpoint (SAC) appears fundamental for taxane mediated cell death. Various mechanisms involved in this checkpoint appear to influence and be influenced by taxanes, and defects in the SAC correlate with resistance. In the favor of this strong interaction are the following data:

1) Activation of the SAC induces both Mad2 and BubR1 interaction with Cdc2 inhibiting its ability to activate APC [55]. Destruction of cyclin B and other regulators of mitosis by APC are responsible for proper metaphase-anaphase transition and mitotic exit. MAPs, including Mad2 and BubR1, regulate SAC, preventing anaphase until chromosomes are attached to bipolar spindles [90]. Spindle inhibitors inhibit cyclin B degradation, cells arrest at pro-metaphase and maintain constitutive Cdk1 activity (destruction of cyclin B inactivates Cdk1) [55];
2) SAC is non-functional when Mad2 levels are low. Sensitivity is restored when Mad2 levels increase [55];
3) Cyclins E and A are important mediators of G1–S phase transition and subsequent S–G2 phase transition. Cyclin A is directly involved in regulating Cdk1 (cdc2) activity as activated Cdk1 is required for cells to enter mitosis and for SAC functionality, both key requirements of taxane sensitivity [91];
4) BRCA1 is also important for SAC control since it regulates BubR1 transcription. BRCA1 is a co-activator of p53 and positively regulates Mad2, thus inhibiting APC activity. Hence, in BRCA1 deficient cells there is a premature onset of anaphase activated by APC by ubiquitination and degradation of cyclin B and subsequent activation of Cdk1 [92] this being linked to paclitaxel resistance;
5) A microtubule regulator, stathmin, destabilizes microtubules by catastrophic promotion and tubulin sequestration. It is active in G2/M transition where Cdk1 inactivates it allowing for M phase entry. Stathmin over-expression has been correlated with resistance to taxanes [93]. In addition, stathmin was profiled as one of 9 proteins differentially expressed in paclitaxel resistant MCF-7 cells [94];

6) Paclitaxel induced apoptosis is inhibited by human epidermal growth factor receptor 2 (HER2) over-expression, because this over-expression transcriptionally up-regulates p21cip1 which associates with p34Cdc2 inhibiting paclitaxel mediated p34 activation delaying cells from entering G2/M and thereby inhibiting apoptosis [95]. HER2 may also directly phosphorylate Cdc2 leading to resistance. It has also been reported that HER2 positive tumors have low levels of Cdk1 resulting in delayed mitosis and paclitaxel resistance [96]. Stimulation through HER2 and other RTKs has also been shown to result in increased levels of P-gp, without affecting transcription, through the MEK–ERK–RSK pathway as inhibitors to this pathway decrease P-gp mediated resistance to paclitaxel [96].

HER family is fundamental for the growth of breast cancer cells. The trans-membrane receptor tyrosine kinase HER2 is over-expressed mainly due to gene amplification in 15%–20% of invasive breast tumors [97-99]. Increased tumor aggressiveness and chemo-resistance to several anti-cancer drugs, including docetaxel and paclitaxel are associated with amplification and/or over-expression of HER2 [100-102]. Breast cancer cell lines with a basal-like gene expression profile, similar to that of triple negative breast cancer (i.e. tumors lacking ER and progesterone receptor (PgR) expression and HER2 amplification), have a greater sensitivity to paclitaxel than HER2-amplified cells [103].

One of the most well known mechanisms relates to drug resistance is associated with the over-expression of the MRD-1 gene product P-gp in cancer [104-106]. Increased expression of this protein has been extensively studied in taxane resistant breast carcinomas. [107]. P-gp is a member of a growing family of at least 49 adenosine triphosphate (ATP) binding cassette (ABC) transporters [108]. Proteins in this family include P-gp/ABCB1, breast cancer resistance protein (BCRP)/ABCG2 and multi-drug resistance related protein (MRP-1)/ABCC1, all of which confer an MDR phenotype [109-112]. The gene for P-gp is localized to chromosome 7, encoding a 170 kDa protein containing two ATP-binding sites and two trans-membrane domains [113]. P-gp expression is correlated with acquired and *de novo* resistance to natural amphipathic products including taxanes, vinca alkaloids, epipodophylotoxins and anthracyclines [104]. As previously mentioned, increased expression of HER2 was associated with a higher resistance to paclitaxel and docetaxel in a panel of established breast cancer cell lines. The cells with over-expression of HER2 also expressed higher levels of P-gp [102]. Most important, neutralizing

HER2 activity by a monoclonal antibody and simultaneous P-gp inactivation by thioridazine or verapamil sensitized cells against paclitaxel in a synergistic manner, indicating that both HER2 and P-gp need to be inhibited to efficiently sensitize cells to paclitaxel, at least *in vitro* [102].

HER2 may also have influence on taxane resistance through interference with survival and apoptosis. Some research showed relevant proteins survivin and p21CIP1 to be associated with both taxane resistance and HER2 amplification/over-expression. Survivin plays an important role in spindle checkpoint regulation during cell cycle progression [114] preferentially during mitosis [115]. In tumors with over-expressed HER2, survivin is mostly over-expressed as well [116, 117]. In addition, in cell lines, HER2 expression positively correlates with survivin expression [116-118].

The heterodimerization of HER2 and HER3 is a mechanism to activate the PI3K/Akt pathway [119, 120]. PI3K/Akt signaling pathway plays a key role in the regulation of cell survival, cell growth and cell cycle progression. This signaling pathway is often excessively activated in malignant tumors leading to decreased sensitivity of tumors to paclitaxel [121]. PTEN tumor suppressor is the major negative regulator of this signaling pathway. The loss of PTEN activity activates PI3K/Akt signaling pathway, leading to increased survival of tumor cells and emergence of resistance to paclitaxel [122]. Upon PI3K activation, Akt is phosphorylated and triggers downstream survival signals, such as the expression of survivin [123, 124]. When inhibitors of PI3K and sarcoma tyrosine kinase (src) are applied, survivin is upregulated resulting in increased apoptosis, which seems to be PI3K-dependent, but Akt-independent [118]. Up-regulation of survivin mediated by HER2 results in paclitaxel resistance due to a faster mitotic exit, mediated by survivin.

Within the HER2 signaling cascade, the major transcription factor related to taxane cytoxicity appears to be Y-Box binding protein 1 (YB-1). In postoperative chemotherapy of breast cancer, YB-1 expression is a negative prognostic factor [125]. In normal breast tissue YB-1 expression is not detectable [126, 127], but it is essential for breast cancer cell growth [128, 129]. Intriguingly, ABCB1 was identified as a target gene of YB-1 binding, supporting its association with drug resistance [126, 130]. One study reported that translocation of YB-1 from the cytoplasm to the nucleus was associated with increased expression of P-gp in breast cancers [131]. The authors claimed that nuclear localization of YB-1 was significantly correlated with resistance to paclitaxel. Their data confirmed previous reports indicating that YB-1 is important in drug sensitivity of cancers via increasing P-gp levels, offering another mechanism for increased P-gp in chemotherapy treated breast cancers.

Inhibition of YB-1 leads to sensitization of breast cancer cell lines to paclitaxel [132], confirming the drug resistance association.

Glutathione S-transferase P1 (GSTP1) conjugates anti-cancer drugs to glutathione, thereby detoxifying cells [133]. An association between GSTP1 protein expression and response to chemotherapy in breast cancer has been often reported [134-136]. Bcl-2 has the ability to suppress apoptosis when linked with glutathione. Namely, it appeared to co-localize with GSTP1 within the nucleus [134]. Inside the nucleus, GSTP1 prevents DNA damage by anti-cancer drugs. Also, less apoptosis was found in tumors being positive for GSTP1 or Bcl-2, as compared to negative counterparts [134]. In MCF7 cells, over-expression of GSTP1 mediates docetaxel resistance [137].

Cytochrome P450 3A4 (CYP3A4) metabolizes docetaxel into an inactive precursor [138]. CYP3A4 mRNA levels are significantly higher in breast tumors with poor response to docetaxel compared to sensitive tumors [139]. Noteworthy, ABCB1-amplified breast cancer cell lines also showed amplification of CYP3A4, both of which reside on chromosome 7q22.1 [140]. Enhanced CYP3A4 expression is also associated with a single nucleotide polymorphism in the 5′-regulatory region of CYP3A4 (−392A>G) [141, 142], reported to increase docetaxel clearance [143].

F-box and WD repeat domain-containing 7 (FBW7) is also involved in the regulation of drug resistance according to report of several groups [144, 35]. The Wertz group reported that FBW7-null colon cancer cells were more resistant to paclitaxel-induced cell death than the wild-type cells [35], as well as the ovarian cancer cell lines with naturally occurring FBW7 mutations [35]. In addition, FBW7 wild-type parental tumors were more sensitive to paclitaxel than tumors with mutant FBW7 [35]. As expected, tumors expressing mutant FBW7 had higher levels of Mcl-1. Wertz et al., proposed that the association between FBW7 and Mcl-1 was disrupted during mitotic arrest due to FBW7 mutations, leading to impaired Mcl-1 degradation and thus resistance to paclitaxel. In addition, inhibition of Mcl-1 in FBW7-null cells restored their sensitivity to paclitaxel-induced cell death, indicating that Mcl-1 is necessary for establishing resistance to paclitaxel in FBW7-deficient cells [35].

One approach to overcome the problem of resistance to paclitaxel is the application of combination therapy, which includes compounds that inhibit signaling pathways for survival or activate additional apoptotic signals. Application of PI3K/Akt signaling pathway inhibitors in combination with paclitaxel significantly increases paclitaxel efficacy in the treatment of tumor cells [145]. Another signaling pathway for survival whose inhibition increases sensitivity of tumor cells to paclitaxel is Raf/MEK/ERK signaling pathway

[146]. In the strategy that involves combined treatment it is not only the antitumor effect of paclitaxel that is increased, but also the inhbition of resistance caused by paclitaxel, which contributes to a favorable therapeutic outcome.

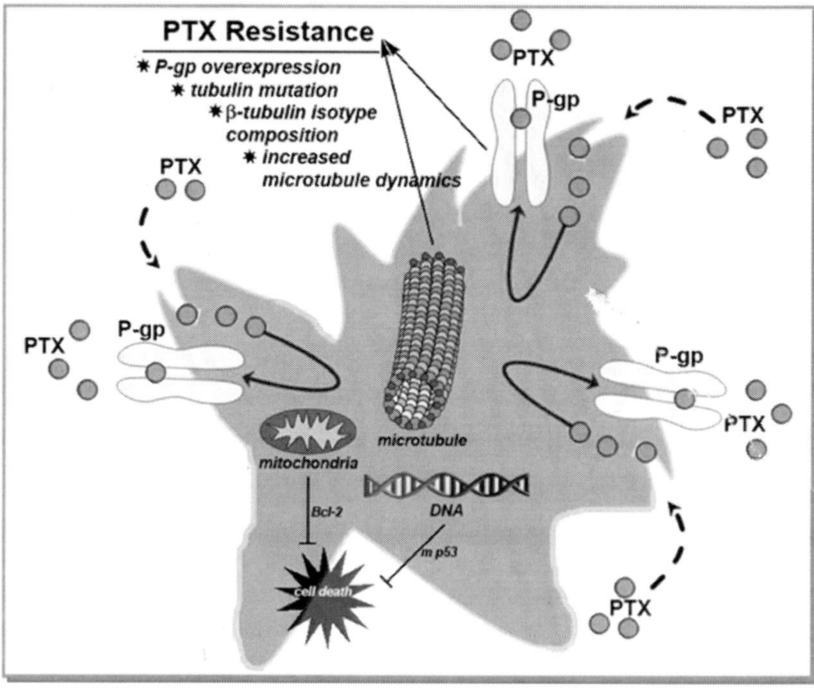

Figure 2. Mechanisms involved in paclitaxel (PTX) resistance. PTX is a substrate for P-gp efflux transporter. Moreover, exposure to PTX could induce the over-expression of P-gp in cancer cells and consequently the development of MDR phenotype. The alterations in PTX target - β-tubulin, such as mutations or prevalent expression of certain isotypes usually contribute to PTX resistance. In addition, resistance to apoptosis due to expression of anti-apoptotic factors (Bcl-2) or inactivation of tumor suppressor p53 could decrease the PTX effectiveness.

Several groups have investigated high throughput screening of thousands of genes as a method to identify patterns of expression of single genes or gene combinations (gene signatures) that correlate with outcome to given therapies. A first study to investigate taxane related outcomes of response in breast cancer was reported by Chang et al. [147]. They identified 92 genes that correlated with docetaxel response of primary breast cancer when applied as neo-adjuvant therapy. The RNA profile from 24 patients included higher expression of genes involved in cell cycle, cell adhesion, protein modification, transcription and apoptosis, while resistant tumors showed increased expression of some transcription factors and signal transduction genes. The 92-gene predictor had positive and negative predictive values of 92% and 83%, respectively. Utilizing another technique, adaptor-tagged competitive (ATAC)-polymerase chain reaction (PCR), Iwao-Koizumi et al., measured the expression of 2453 genes in a series of 70 (44 learning set, 26 validation set) primary or locally recurrent breast cancers receiving docetaxel [137]. They identified an algorithm consisting of 85 genes that predicted clinical response to docetaxel with positive and negative predictive values of 73.3% and 90.9%, respectively. Non-responders were characterized as having elevated expression of genes controlling cellular redox, thioredoxin, glutathione-S transferase and peroxiredoxin.

Nevertheless, the most studied mechanisms of paclitaxel resistance, whose overcoming represents the starting point for the development of more effective drugs, are the P-gp over-expression, tubulin mutations and alterations in microtubule composition (Figure 2).

New Taxanes

In last decade, considerable efforts have been made in the field of microtubule-targeting chemotherapies discovery in order to increase the selectivity, efficacy, central nervous system access, as well as to decrease toxicity and the development of drug resistance observed with the most promising anti-cancer agents – paclitaxel and docetaxel (Figure 3) [148]. A number of taxanes side effects have been associated with the solvents used for their dilution: CrEL2 for paclitaxel and polysorbate 80 (Tween 80) for docetaxel. These side effects include the acute hypersensitivity infusion reactions and peripheral neuropathy. In addition, these solvents could change paclitaxel and docetaxel pharmacokinetic profiles [149]. Many different strategies were developed to create new formulations of paclitaxel and docetaxel. Here are the examples:

Modification with a nano-particle albumin-bound paclitaxel (Abraxane) made a colloidal suspension, which enhanced the intracellular delivery of paclitaxel and consequently its activity [150, 151]. Compared to conventional paclitaxel, Abraxane was associated with lower incidence of side effects [148]. Nab-docetaxel, the nano-particle albumin bound docetaxel formulation related to Abraxane, was also developed [152] (Table 1).

Cabazitaxel is a semi-synthetic dimethyloxy derivative of docetaxel (Table 1). The alteration in the structure of the docetaxel backbone achieved by substitution of the hydroxyl groups with the dimethyloxy side chains caused lower P-gp affinity. Furthermore, the presence of the extra methyloxy side chains could increase the ability of cabazitaxel to cross the blood–brain barrier [153].

Tesetaxel is an orally bio-available semi-synthetic taxane derivative with high solubility, lack of neurotoxicity and impressive anti-tumor activity (Table 1). Unlike other taxanes, the efficacy of tesetaxel was unaffected by the P-gp expression levels. In addition, the intracellular accumulation of tesetaxel was much higher than that of paclitaxel or docetaxel, particularly in P-gp-positive cells [154].

Paclitaxel poliglumex (PPX) is a novel biodegradable polymeric drug conjugate of paclitaxel with α-poly-L-glutamic acid (Table 1). It is designed to enhance the solubility and permeability of paclitaxel, to minimize normal tissue exposure to free drug, and to avoid the development of multi-drug resistance (MDR). This novel formulation of paclitaxel does not contain CrEL and therefore pre-medication with steroids and antihistamines is not required [155].

Cationic liposomal paclitaxel or EndoTAG-1, which does not contain CrEL, was designed to improve efficacy and toxicity profile of CrEL–paclitaxel (Table 1). In addition, these cationic liposomes target tumor endothelial cells. EndoTAG-1 was shown to affect tumor microvasculature by causing functional impairment, tumor-selective vessels occlusion, and micro-vessel leakiness. These anti-angiogenic characteristics give additional advantage of EndoTAG-1 over paclitaxel [156].

Larotaxel is a semi-synthetic taxoid derived from 10-deacetyl baccatin III as a major natural compound of the yew tree needles (Table 1). It is also a tubulin-targeting drug. The advantages of larotaxel are its ability to cross the blood–brain barrier [157] and its activity in both taxane-sensitive and -resistant cell lines.

Polymeric-micellar paclitaxel is paclitaxel with a biodegradable polymeric-micellar nano-particle [158]. Theoretically, the copolymer residue increases the water solubility of the hydrophobic paclitaxel and permits delivery of high doses of paclitaxel. It was shown to be a more potent radio-sensitizing agent than paclitaxel [159].

A natural fatty acid, docosahexaenoic acid (DHA) was conjugated through an ester bond to the paclitaxel 2′-oxygen (DHA-paclitaxel). The idea for this conjugation was that tumors would readily take and use the natural fatty acids as energy sources [160]. DHA-paclitaxel was more potent than paclitaxel in tubulin polymerization and more active against typically taxane resistant tumors [148]. However, it is no longer in clinical trials.

ANG1005 is another modified form of paclitaxel that circumvents the use of cremophor CrEL2 (Table 1). ANG1005 consists of paclitaxel molecules conjugated to a receptor-targeting peptide that allows selective transport across the blood–brain barrier. Upon entry into cells, ANG1005 undergoes esterase cleavage to release three paclitaxel molecules from each receptor peptide [161].

Other taxanes were also designed to overcome acquired resistance to taxane-based therapies by circumventing the P-gp drug efflux: TPI-287, BMS-188797, Ortataxel and Milataxel [162] (Table 1).

Many of these new taxane formulations were withdrawn from clinical studies due to different reasons. Nevertheless, the discovery of new taxanes with microtubule-stabilizing activity still comes forward.

Resistance to taxanes induced by P-gp and other MDR efflux pumps is one of the main obstacles to successful chemotherapy. Therefore, different strategies for blocking the extrusion of drugs have been proposed [163]:

Direct inhibition of transporters;
Exploit the collateral sensitivity of MDR cells to certain cytotoxic agents;
Engagement of cytotoxic agents that are not substrates for MDR proteins.

MDR modulators are actually MDR transporter inhibitors that are often co-administered with classic cytotoxic agents. A number of different chemical classes including some second-generation taxanes are recognized as MDR modulators [164].

Verapamil, a calcium channel blocker was among the first generation of MDR modulators [165]. Most of first generation MDR modulators were known as therapeutics agents already used for other indications. They exerted the MDR reversal potential at concentrations that were toxic. Further investigations led to the development of less toxic second-generation modulators and afterwards third-generation modulators active in nano-molar range with fewer pharmacokinetic interactions. However, despite the promising advances in preclinical models, to date clinical studies with MDR modulators had limited success [163].

Table 1. New microtubule-interacting drugs in clinical trials and their advantages over paclitaxel according to http://clinicaltrials.gov and Dumontet and Jordan, 2009 [197]

New microtubule-interacting drugs	Clinical trials	Crosses blood–brain barrier	Interaction with P-gp/MDR modulator	Efficient in spite of tubulin alterations
Abraxane (nano-particle albumin-bound paclitaxel)	Breast cancer	No	No/Radio-sensitizing properties	No
Nab-docetaxel	Prostate and breast cancer	Yes	No/Radio-sensitizing properties	No
Cabazitaxel	Prostate cancer	Yes	Low	No
Tesetaxel	Various types of cancer, including breast, gastric and other solid tumors	/	No	No
Paclitaxel poliglumex (PPX)	Colorectal, brain, esophageal and prostate neoplasms	Yes	No	No
EndoTAG-1 (cationic liposomal paclitaxel)	Breast, liver, pancreas cancer, metastases	/	No	No
Larotaxel	Breast, pancreatic, urinary bladder neoplasms, metastases	Yes	Poor	No
ANG1005	Malignant glioma, brain metastases	Yes	No	No
TPI-287	Neuroblastoma, medulloblastoma, glioblastoma multiforma, prostate cancer	Yes	No	No
Ortataxel	Taxane-resistant tumors	/	No	No
MAC-321 (Milataxel)	Colorectal and non-small cell lung cancer, mesothelioma	/	No	Yes
Epothilone B (Patupilone)	Brain metastases in breast and ovarian cancer, melanoma and other solid tumours	Yes	No/Modulator	Yes
Ixabepilone	Breast cancer and other solid tumors	/	No	Yes
Sagopilone (ZK-EPO)	Glioblastoma, prostate and lung cancers	Yes	/	Yes
KOS-1584	Non-small cell lung carcinoma, breast and prostate cancer	Yes	/	Yes
Discodermolide	Trial terminated in phase I	/	No	Yes
Eribulin	Advanced metastatic breast cancer	Yes	No	Yes

Hydrophobic peptides, which interact with trans-membrane segments of P-gp and interfere with the proper assembly and functioning of the protein, may also reverse the MDR [166].

Blocking the expression of *mdr1* mRNA (codes for P-gp) by antisense oligonucleotides [167] and preventing the P-gp synthesis using chemical compounds [168, 169] are among engaged strategies for overcoming the MDR.

An interesting approach is based on the idea that drug efflux pumps can be exploited by certain drugs in order to selectively kill resistant cancer cells. The co-administration of a cytoprotective agent, which is a substrate for efflux pumps, with a cytotoxic agent, which is not recognized by MDR transporters leads to the efficient exposure of resistant cells to cytotoxic therapy, while normal cells stay protected and unharmed [170]. In addition, the use of anti-P-gp antibodies may result in selective killing of MDR cells [171]. Recent findings imply that cationic naturally occurring antimicrobial peptides may selectively kill P-gp over-expressing MDR cells [172].

Another approach to overcome resistance is evading strategy based on the development of drugs, which are not substrates for MDR transporters [173]. Interestingly, it was found that chemical modifications of paclitaxel and docetaxel can produce active second generation taxanes that are not transported by MDR pumps [5].

During the last two decades, approximately 120 taxanes with different skeletons, have been isolated from the Japanese yew, *Taxus cuspidata*. Some of these agents were able not only to exert significant cytotoxic activity, but also to reduce Ca2+-induced depolymerization of microtubules and to increase cellular accumulation of vincristine in MDR cancer cells [174]. Taxinine NN-1 (taxezopidine G), cyclotaxinine NN-2, taxuspine X and taxuspine C showed significant reversal activity due to increased vincristine accumulation in MDR cells [175, 174]. Plant cell cultures of *Taxus* species were used as a biotechnological approach aimed to provide paclitaxel and related taxanes in amounts necessary for clinical application. In that way, taxinine NN-11 was isolated from callus culture of *Taxus cuspidata* [176]. Taxinine NN-11 displayed higher activity than verapamil towards vincristine accumulation. Further chemical investigation of *Taxus cuspidata* callus cultures led to the isolation of the new taxane – the 9,10-isomer of taxinine NN-11 [177]. Taxusin was isolated as a secondary metabolite of the dark brown callus culture of *Taxus cuspidata* incubated under light irradiation. This taxane also exhibited stronger MDR reversal activity than verapamil [178].

New natural taxanes showed significant anti-P-gp potential and stimulated continuous search for new natural or synthetic paclitaxel derivatives with improved MDR reversal activity.

Broad structure-activity relationship studies of synthetic taxanes led to the development of highly efficient taxane-based MDR reversal agents and cytotoxic taxanes with considerable ability to evade the MDR transporters (second-generation taxanes). Some of these newly synthesized taxanes are in different phases of clinical development.

Three novel taxinine analogs were prepared from sinenxan A (a biosynthetically available taxane) [179], while a series of taxuspine C analogs were synthesized and their effect on the cellular accumulation of vincristine was examined [180]. Taxinine derivatives containing a cinnamoyloxy or a benzoyloxy group at C2, C5, or C13 significantly increased the cellular accumulation of vincristine suggesting their good MDR modulating activity. Recently, the synthesis of a series of 2-deacetoxytaxinine J derivatives was reported and the best results on paclitaxel accumulation were achieved with a taxinine bearing a benzoyl group at C13 position [181].

It seems that phenyl ring of paclitaxel and many P-gp substrates is lacking in second-generation taxanes, suggesting that this may be a reason for their poor transport by P-gp. The markedly higher activity of second-generation taxanes than paclitaxel *in vitro*, especially in P-gp highly expressing cancer cell lines suggests the clinical potential of second-generation taxanes. Second-generation taxanes: SB-T-1102, 1214 and 1216 [182] were more efficient than paclitaxel in highly P-gp-positive cells, eventually due to the lack of their P-gp transport.

High CYP-catalyzed metabolism almost suppressed the cytotoxicity of paclitaxel *in vitro* [183]. In addition, it was observed that IDN5390 increased Cyp3a1 expression, which might be a reason for its inactivity *in vivo*. In contrast, the effective IDN5109 decreased Cyp3a1 expression. CYP3A enzymes metabolize SB-T-12854 (fluorinated taxane) to products that are even more efficient. An important role of Cyp3a1 expression on new taxane activity especially *in vivo* cannot be excluded when we bear in mind that CYP enzymes are tightly connected with P-gp and strongly influence pharmacokinetics of drugs and their availability in the organism [184].

Paclitaxel exerts its anticancer effect partly by inhibiting growth factors and tumor angiogenesis [185]. These properties were also recognized for new taxanes- IDN5109 and IDN5390 [186, 187]. Vascular endothelial growth factor (VEGF) is the most important pro-angiogenic factor that acts as mitogen for vascular endothelium. Its binding to the tyrosine-kinase receptors in the membrane initiates a cascade of intracellular signals resulting in endothelial cell proliferation and the formation of new blood vessels [188, 189]. Recently, we have also shown that paclitaxel inhibits VEGF secretion. Moreover, this effect was observed in paclitaxel-sensitive as well as in paclitaxel-resistant cancer cell lines [190]. The suppressed expression of tumor angiogenesis related genes by new taxanes [184] indicates that their therapeutic efficiency is not limited to highly expressing P-gp cancer cells, but also involves the inhibition of other mechanisms important for tumor progression.

Taxanes that entered clinical trials were more effective in tubulin-mutant cell lines in comparison with paclitaxel and showed improved pharmacokinetic properties. However, reduced, but still significant resistance to BMS-184476, BMS-188797 and BMS-275183 [191] was found in the paclitaxel-resistant cells harboring point mutations in β-tubulin [192]. Considerably lower levels of cross-resistance were observed for taxanes MAC-321 and MST-997 when compared with paclitaxel and docetaxel in cancer cell lines with β-tubulin mutations [193-195]. Since new taxane – IDN5390 bound strongly to class III β-tubulin whereas paclitaxel did not [196], IDN5390 was more effective than paclitaxel in paclitaxel-resistant cells. It has been hypothesized that these taxanes interact with tubulin in a binding site similar but distinct from that of most taxanes [193, 195], or that they interact more efficiently with the mutant tubulin [194].

Although β-tubulin mutations contribute to the paclitaxel resistance *in vitro*, there is no certain evidence that they are important for the development of clinical resistance [72].

MDR Reversal Strategies with Microtubule-Stabilizing Agents

Chemotherapy often fails to cure cancer patients due to unexpected drug resistance. Increasing drug sensitivity is now the imperative and many novel approaches for successful treatment were designed [198]. There are several reasons for the acquisition of drug resistance [199, 200]:

1) As mentioned before, cancer cells may increase the expression of energy-dependent ATP-binding cassette (ABC) transporters that efflux the anti-cancer drugs out of the cells [200]. ABC drug transporters particularly involved in the extrusion of many clinically important drugs are ABCB1 (P-glycoprotein - P-gp), ABCC1 (MDR-

associated protein 1 - MRP1), and ABCG2 (breast cancer resistant protein - BCRP) [200, 201];

2) Impaired function of solute carriers that provide drug influx may lead to the decreased concentration of drug in cancer cells [200];
3) Increase in DNA repairing efficiency may cause the insensitivity to chemotherapeutic drugs [200];
4) Recruiting of drug-detoxification mechanisms may result in the resistance to drug-induced apoptosis [202];
5) Change in cellular polarity and behavior, i. e. epithelial-mesenchymal transition (EMT) is associated with acquired drug resistance [203];
6) Alterations in the expression or function of specific genes involved in the control of cell growth, apoptosis, cell cycle regulation, and invasion may contribute to drug resistance [204];
7) Recently, it was shown that cancer stem cells (CSCs) are involved in drug resistance in various types of human cancers such as breast cancer, colorectal cancer, leukemia, glioblastoma, small cell lung carcinoma, ovarian cancer, prostate cancer, and pancreatic cancer [205]. However, it was observed that CSCs resist chemotherapy due to increased expression of ABCG2 and ABCC1, which protect CSCs from chemotherapeutic assault [206].

Despite the huge knowledge about drug resistance mechanisms, physicians often do not recognized and understand the clinical relevance of any of the aforementioned mechanisms of resistance. This maybe explained by the lack of universally accepted guidelines for clinical validation, differences in methods of tissue collection preparation and storage, utilization of different target assays and the problems in sensitivity and specificity of currently used immuno-histochemical analyses [207, 208].

For example, sample and analytical heterogeneity significantly limited the collection of consistent data for P-gp expression [209]. Unfortunately, the information about P-gp expression in patients chosen for paclitaxel (and many other drugs) treatment is necessary for the prediction of their response to therapy. Besides P-gp, biomarkers of resistance to paclitaxel should have some functional interaction with micotubule assembly, cell cycle disturbance and apoptosis. Considering that taxanes do not directly cause DNA-damage and that the induced cell cycle arrest occurs in a late phase of G2/M, there is little supposed benefit from the study of p53 status. Mutational analysis of p53 is varying frustrating and with errors, because we do not know exactly what each mutant variant does. Some generate stable aberrant protein, and others generate no protein at all [210]. Furthermore, some mutation positive cases may generate mutant p53 that is undetectable by a given antibody clone used for IHC.

The only promising predictive clue is the positive correlation between taxane efficacy (it is worth for many other chemotherapeutics) and the proliferation index of the tumor [211-213].

Combinatorial therapies give an exciting opportunity for the development of new chemotherapy strategies [214]. The synergistic effects of drug combinations are able to reduce the concentrations of two drugs diminishing the unwanted side effects of either drug used alone. A beneficial clinical impact of drugs with distinct mechanisms of action and non-overlapping toxicities is the primary principle for the development of combiantion therapies. Currently, taxanes are now used in various combinations with the platinum reagents cisplatin and carboplatin and the nucleoside analog gemcitabine in the treatment of non-small cell lung cancer, and the anthracyclines doxorubicin and epirubicin in breast cancer treatment [215-218]. All of these agents interfere with DNA. While platinum reagents induce direct damage, gemcitabine incorporates into DNA in place of deoxycytidine, resulting in chain termination and the inhibition of synthesis [218, 220]. The anthracyclines promote DNA double-strand breaks and the inhibition of topoisomerase II [221]. The benefits achieved from current combinations are encouraging, but a greater understanding of the molecular effects of microtubule-stabilizing agents could facilitate the discovery of more effective therapeutic combinations.

Despite significant medical advances, cancer persists to be incurable disease. New therapeutic options are a high priority for many pharmaceutical companies and independent research organizations worldwide. Current drug discovery involves screening for potential lead compounds with specific targets and fewer side effects. Natural products are still the most promising anti-cancer agents even though *de novo* design and synthesis of bioactive molecules with sophisticated computer modeling are now available. Modern science is unlikely to replace the millions of years of evolution and natural selection that exist in living organisms [222, 223].

The epithiliones, a family of naturally occurring cytotoxic microtubule inhibitors, structurally different from paclitaxel and docetaxel, represent the most advanced agents for taxane resistant or refractory patients. They were named for their molecular features (epoxide, thiazole, and ketone) [224]. This class of substances comprises naturally occurring epothilones A–F and synthetic derivatives such as ixabepilone, ZK-epothilone, and KOS-1584

(9,10,-didehydroepothilone D). Epothilones A and B (Figure 3) were originally isolated as anti-fungal fermentation products of the myxobacteria Sorangium cellulosum [225]. Shortly thereafter, these agents were demonstrated to stabilize microtubules and induce mitotic arrest in G2/M resulting in apoptosis similar to that of the effect of taxanes [226]. Although they bind to similar sites as the taxanes, they have a unique and independent molecular interaction that is partly to explain their resistance to the classical MDR-drug resistance mechanisms [227]. A crystal structure of their interaction with microtubules has not yet resolved, but it is known that epothilones compete with paclitaxel for microtubule binding [228]. Nevertheless, their cytotoxic activity is the consequence of the same mechanism, but the epothilones have significant advantages over taxanes:

1) They originated from myxo-bacteria, which make them available in large-scale culture and isolation. They are also easier to manipulate. Chemical syntheses, either total or partial, of these molecules and their analogs allows modification in order to enhance their efficacy [229];
2) They have a unique pharmacophore that binds at or near the taxane-binding site, resulting in a similar but slightly distinct mechanism of action [230];
3) An attractive feature of the epothilones is that they are poor substrates for P-gp mediated transport, which make them efficient in a considerable subset of taxane-resistant tumors [231].

The epothilones were quickly recognized as effective against cell lines expressing P-gp and there are direct evidences for P-gp incapability to react with epothilones. Sensitivity to paclitaxel-induced cytotoxicity in several paclitaxel-resistant lines was restored in the presence of P-gp inhibitors such as verapamil, while sensitivity to an epothilone was unaffected. Moreover, paclitaxel-sensitive lung carcinoma cells became resistant to paclitaxel when stably transfected with a construct encoding P-gp, but these cells remained sensitive to epothilone-mediated death [228, 233]

The antitumor activity of the natural epothilones and the ability to overcome mechanisms of resistance led to the development of new epothilone derivatives. Several epothilones or epothilone-derivatives that were effective against cancer cell lines and tumor xenografts are now in clinical trials [234]. The lactam analogue of epothilone B, ixabepilone, was designed to enhance metabolic stability to plasma esterases [235]. The loss of the epoxide moiety in epothilone D (KOS-862) and its analogue KOS-1584 did not affect the biological potency of these compounds [236, 237].

These epothilone derivatives are in clinical development with over 100 separate trials in progress against solid tumors and lymphoma, both as single agent therapy and in combination with other agents. Ixabepilone (Ixempra ; Bristol-Meyers Squibb), was approved in the fall of 2007 for metastatic or locally advanced taxane- and anthracycline-resistant breast cancer [238]. Phase II studies examining the effect of ixabepilone in chemotherapy-resistant lymphoma are also ongoing. Other epothilones progressing through clinical development include epothilone B (Patupilone), ZK-EPO (Sagopilone; Bayer) and KOS-1584 (Kosan Biosciences) (http://clinicaltrials.gov) (Table 1).

Ixabepilone is the most developed of epothilone derivatives with low resistance (*in vivo* and *in vitro*) to various drug resistance mechanisms that affect the taxanes including P-gp over-expression, β-tubulin mutations, β-III isotype over-expression (Table 1). Moreover, ixabepilone weekly induces P-gp expression and retains the ability to bind to β-III microtubules [239]. Based on preclinical data supporting its efficacy in paclitaxel-sensitive and paclitaxel-resistant cell lines, the clinical development of ixabepilone in breast cancer has focused on patients with tumors that had failed prior treatment with anthracyclines and/or taxanes [235, 240]. As a single agent or in combination, ixabepilone possesses manageable safety profile. Currently, it is used in combination with capecitabine for the treatment of metastatic breast cancer resistant to anthracycline and taxane or as a monotherapy for metastatic breast cancer refractory to anthracyclines, taxanes, and capecitabine. Ongoing studies in combination with targeted agents include a large phase II trial of trastuzumab plus ixabepilone or docetaxel in HER- 2–positive metastatic breast cancer patients and a phase II trial of ixabepilone plus cetuximab as first-line therapy for triple-negative metastatic breast cancer patients (http://clinicaltrials.gov). Active trials also include phase II studies of ixabepilone plus bevacizumab versus paclitaxel plus bevacizumab as first-line therapy for metastatic breast cancer. Ixabepilone and carboplatin with or without bevacizumab were studied in advanced non-small cell lung carcinoma, as well as ixabepilone plus cetuximab as firstline therapy for metastatic pancreatic cancer (http://clinicaltrials.gov).

Epothilone B, ZK-EPO, and KOS-1584 were investigated in patients with advanced solid tumors, including gastric, hepatocellular, ovarian, prostate, and renal cancers (Table 1). An important property of epothilone B is its ability to cross the blood–brain barrier. Consistent with these results, epothilone B has demonstrated activity against recurrent/progressive metastatic lesions to the brain in patients with non-small cell lung carcinoma and metastatic breast cancer [241, 242]. Ongoing trials are evaluating epothilone B in platinum refractory ovarian, fallopian, and peritoneal cancers, glioblastoma, and prostate cancer (http://clinicaltrials.gov).

Further steps in the development of epothilone derivatives are the exploration of combinations with classic chemotherapeutics as well as targeted agents, evaluation of their activity during earlier stages of disease, and identification of predictive markers to guide therapy.

Spurges are worldwide spread plants suggested to be fruitful producers of structurally unique macrocyclic diterpenes of which jatrophane type is prominent. Several new jatrophanes isolated from spurges (*Euphorbiaceae*) were shown to be potent MDR modulators [243-246]. It was reported that some jatrophanes exhibit microtubule-interacting activity similar to paclitaxel, but without inducing cell cycle arrest in G2/M phase [247]. We have also confirmed that jatrphanes stimulate purified tubulin assembly *in vitro* achieving microtubule-interacting activity with concentrations that effectively inhibit the cell growth [248]. Although the exact mechanism of jatrophanes-mediated cell cycle disturbance needs to be clarified, we showed that the synergism between jatrophanes and paclitaxel is partly due to their mutual effect on microtubule assembly [248]. Since macrocyclic diterpenes of jatrophane type previously isolated from *E. dendroides* were described as P-gp pump inhibitors [249], we assumed that our jatrophanes (euphodendrophanes, Figure 3) originated from the same plant share common features. Indeed, a marked increase in the accumulation of P-gp substrate Rho123 was observed in MDR cancer cell line - NCI-H460/R [250] treated with jatrophanes. Moreover, the results implicated that jatrophanes were more efficacious at P-gp inhibition then Dex-verapamil, second-generation MDR modulator [248, 246]. We have further examined the anti-cancer potential of jatrophanes and found that they inhibit the secretion of VEGF. This effect was more pronounced in combination with paclitaxel [248].

An unexpected source for the identification of microtubule stabilizing agents has been marine organisms. The main feature of marine life is the huge diversity that is due to the enormous space covered with oceans as well as to the ancient origin of marine organisms, which evolved hundreds of millions of years resulting in great evolutionary variation of the marine species [251, 252]. Marine-derived drugs are generally more potent than plant-derived compounds because of the dilution effect of the ocean environment. The success of marine-derived drugs in the clinical treatment of cancer has been particularly encouraging [253]. Sponges have proven to be one of the most successful groups of the many different types of marine organisms used as a source for drug discovery [254]. Marine sponges anchorage microorganisms on their surfaces, in their canal systems, and in their intercellular spaces. Most of these microorganisms are endosymbionts and are probably the major source of the compounds isolated from marine sponges [255]. Some indirect evidence for this is the fact that the production of a certain compound often vary between individual sponges, which correlates with their locations, ranging from some that produce no compound, to others that produce large amounts of the compound [256].

Laulimalide and isolaulimalide (Figure 3) are natural products of the marine sponge *Cacospongia mycofijiensis* [257]. Laulimalide (fijianolide) is a microtubule stabilizer that has a novel microtubule-binding site on tubulin, which allows its synergism with the taxanes [258]. Laulimalide binds to a different site to that of the classic microtubule-stabilazing agents (paclitaxel, epothilone, and discodermolide), which all bind to the well-described taxoid site on β-tubulin and therefore compete for each other's binding [259, 260]. Additional advantages of laulimalide include efficacy in P-gp-expressing cell lines and potential antiangiogenic activity [261, 257]. The main obstacle for its usage is the enormous difficulty to synthesize the sufficient quantities for treatment and substantial instability. In contrast to its unquestionable efficacy in cancer cell lines, laulimalide demonstrated controversial results *in vivo*, even inducing the severe toxicity in some cases [262, 263]. Therefore, a number of simplified laulimalide analogs have been synthesized that retain the anti-mitotic activity of laulimalide while possessing increased stability [264].

Eleutherobin (Figure 3) was originally isolated from the Western Australian Alcyonacean coral *Eleutherobia* sp. (possibly *Eleutherobia albiflora*) and shown to cause mitotic arrest by the stabilization of microtubules. Eleutherobin and its derivatives belong to a small but highly significant family of diterpene glycosides. These are highly potent cytotoxic agents with activity comparable to paclitaxel. Unfortunately, the lack of supply has greatly limited the development of eleutherobin as a prospective anti-cancer agent and thus its full potential is unknown. Eleutherobin has not been followed clinically, probable due to its susceptibility to P-gp-mediated transport [265]. This family of highly promising diterpene glycosides was enlarged with the isolation of six new eleutherobin analogues from the relatively common Caribbean gorgonian *Erythropodium caribaeorum*. Desmethyleleutherobin (IC50 = 20 nM) and isoeleutherobin A (IC50 = 50 nM) were both slightly more potent than eleutherobin (IC50 = 100 nM).

The most interesting marine product is discodermolide (Figure 3). It was initially isolated from extracts of the sponge *Discodermia dissoluta* and identified as an immunosuppressive agent. Now discodermolide is synthetically produced and modified to create novel analogs [266, 267]. It is substantially more potent than paclitaxel in most cell lines, including those expressing P-gp [268, 269]. Like epothilones and eleuthorubin, discodermolide competes with paclitaxel for binding to tubulin. Unlike other microtubule stabilizers, its effect on microtubules exhibits different properties. Microtubule nucleation induced by discodermolide is immediate leading to shorter microtubules than those induced by paclitaxel [230, 270]. The combination of paclitaxel with discodermolide induces synergistic effect while this was not observed in combinations of paclitaxel with epothilones or eleutherobin. Those findings suggest that the binding site of discodermolide may differ from that of paclitaxel. Discodermolide is also able to induce cell killing in paclitaxel-resistant cancer cell lines bearing β-tubulin mutations [230, 271].

Discodermolide has entered phase I trials because of promising preclinical data. Optimal dosing conditions with favorable pharmacokinetics were found in phase I clinical studies. However, an unanticipated side effect of severe pulmonary toxicity was observed and further clinical development was suspended [272] (Table 1). The limited natural supply and difficult synthesis have prevented *in vivo* analysis of dictyostatin, which appears to be a cyclic analog of and share microtubule contacts with discodermolide [273, 274].

Peloruside A (Figure 3) is a novel secondary metabolite isolated from a New Zealand marine sponge, *Mycale hentscheli* that has potent paclitaxel-like microtubule-stabilizing activity at nanomolar concentrations. Its 16-membered macrolide ring is similar to that of epothilone. Like paclitaxel, peloruside A arrests cells in the G_2/M phase of the cell cycle and induces apoptosis. Further investigations determined that peloruside A stabilized the polymerized form of tubulin and induced microtubule bundling in interphase cells and multiple asters in mitotic cells in a similar manner to paclitaxel [275]. The relatively simple structure of peloruside A makes it suitable for the design and synthesis of analogs with improved tumor targeting and reduced tumor cross-resistance. Peloruside A shares many of the same properties of laulimalide, including its binding site and synergistic effects with the taxanes [276, 277].

Obviously, the mechanism of microtubule stabilization by peloruside A differs from stabilization by taxoid-site drugs [278]. Taxoid-site drugs interact with the βH7 region (bottom of the taxoid-binding cleft) that lies next to the T5 loop at the exchangeable nucleotide-binding region, whereas, peloruside A interacts with the H9-S8 loop at the proposed binding site of peloruside A [278]. This binding site also lies next to the T5 loop. A conformational change in the T5 loop is believed to promote improved interactions across the interdimer interface, causing stabilization.

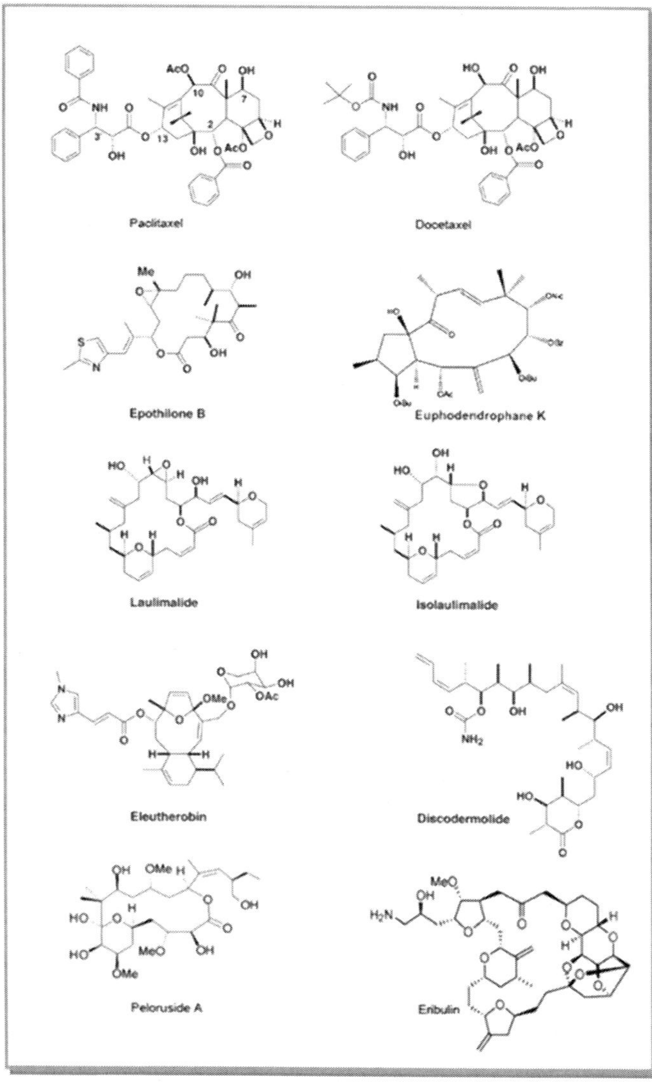

Figure 3. Chemical structure of paclitaxel, docetaxel and functionally related microtubule-interacting compounds.

Peloruside A has some promising advantages over paclitaxel, being more soluble and therefore not requiring the use of Cremophore EL. This should correlate with fewer vehicle-associated side effects than paclitaxel [279]. Peloruside A is also more effective against cells that acquired the MDR-drug resistance phenotype, since peloruside A remains active in the presence of P-gp [230]. Tests of peloruside A in animals have been very promising, with peloruside A showing no explicit toxicity in mice and being more efficacious in inhibiting tumor growth than paclitaxel and docetaxel [280].

Eribulin (Figure 3), a synthetic analog of the marine macrolide halichondrin B, is suggested to be a microtubule inhibitor with a unique mechanism of action. Halichondrin B is a naturally occurring compound originally isolated from the marine sponge *Halichondria okadai* by Hirata and Uemura in 1986 [281]. The complete synthesis of halichondrin B has stimulated the development of structurally simplified and pharmaceutically optimized analog eribulin. Eribulin was approved for the treatment of patients with metastatic breast cancer who have received at least two prior chemotherapy regimens for late-stage disease, including both anthracycline- and taxane-based chemotherapies (Table 1). Eribulin inhibits microtubule stability by blocking microtubule growth without affecting microtubule shortening, thereby sequestering β-tubulin into nonfunctional aggregates and leading to formation of abnormal mitotic spindles and ultimately apoptosis [282, 283].

Conclusion

Taxanes are likely to remain a standard first-choice treatment for many solid malignancies in the near future. However, drug resistance often compromises the clinical benefits of taxanes. Cellular resistance to taxanes is a very complex phenomenon involving different molecular mechanisms, such as over-expression of MDR efflux pumps affecting intracellular drug accumulation, alterations of tubulin inducing a reduced sensitivity to the drugs, alterations in the signaling pathways, in the cell cycle, and in the control of apoptosis. A better understanding of the mechanisms involved in taxane resistance is expected to provide useful tools for the rational design of novel microtubule-stabilizing agents able to overcome simultaneously different resistance phenotypes.

The use of other microtubule-targeting agents may bring clinical benefit to patients with taxane-resistant tumors who otherwise have very few therapeutic options. Several emerging novel microtubule targeting agents, taxane-derivatives or other naturally occurring drugs, especially from marine species, appear to have significant therapeutic potential. The development of these microtubule inhibitors is an imperative for helping the patients with the progressing disease after standard treatment with anthracycline- and taxane-containing regimens.

References

[1] McGrogan, BT; Gilmartin, B; Carney, DN; McCann, A. Taxanes, microtubules and chemoresistant breast cancer. *Biochim Biophys Acta,* 2008 1785, 96-132.

[2] Ojima, I; Wang, T; Miller, ML; Lin, S; Borella, CP; Geng, X; Pera, P; Bernacki, RJ. Synthesis and structure-activity relationships of new second-generation taxoids. *Bioorg Med Chem Lett,* 1999 9, 3423-3428.

[3] Yvon, AM; Wadsworth, P; Jordan, MA. Taxol suppresses dynamics of individual microtubules in living human tumor cells. Mol Biol Cell, 1999 10, 947-959.

[4] Blagosklonny, MV; Fojo, T. Molecular effects of paclitaxel: myths and reality (a critical review). *Int J Cancer,* 1999 83, 151-156.

[5] Galletti, E; Magnani, M; Renzulli, ML; Botta, M. Paclitaxel and docetaxel resistance: molecular mechanisms and development of new generation taxanes. *ChemMedChem,* 2007 2, 920-942.

[6] Wani, MC; Taylor, HL; Wall, ME; Coggon, P; McPhail, AT. Plant antitumor agents. VI. The isolation and structure of taxol, a novel antileukemic and antitumor agent from Taxus brevifolia. *J Am Chem Soc,* 1971 93, 2325-2327.

[7] Guénard, D; Guéritte-Voegelein, F; Dubois, J; Potier, P. Structure-activity relationships of Taxol and Taxotere analogues. *J Natl Cancer Inst Monogr*, 1993 15, 79-82.

[8] Rowinsky, EK. The development and clinical utility of the taxane class of antimicrotubule chemotherapy agents. *Annu Rev Med,* 1997 48, 353-374.

[9] Chan, OT; Yang, LX. The immunological effects of taxanes. *Cancer Immunol Immunother,* 2000 49, 181-185.

[10] Bergstralh, DT; Ting, JP. Microtubule stabilizing agents: their molecular signaling consequences and the potential for enhancement by drug combination. *Cancer Treat Rev,* 2006 32, 166-179.

[11] Goda, K; Bacsó, Z; Szabó, G. Multidrug resistance through the spectacle of P-glycoprotein. *Curr Cancer Drug Targets,* 2009 9, 281-297.

[12] Wu, CP; Hsieh, CH; Wu, YS. The emergence of drug transporter-mediated multidrug resistance to cancer chemotherapy. *Mol Pharm,* 2011 8, 1996-2011.

[13] Fojo, T; Menefee, M. Mechanisms of multidrug resistance: the potential role of microtubule-stabilizing agents. *Ann Oncol,* 2007 Suppl 5, v3-v8.

[14] Chien, AJ; Moasser, MM. Cellular mechanisms of resistance to anthracyclines and taxanes in cancer: intrinsic and acquired. *Semin Oncol,* 2008 35 (2 Suppl 2), S1-S14; quiz S39.

[15] Kamath, K; Wilson, L; Cabral, F; Jordan, MA. BetaIII-tubulin induces paclitaxel resistance in association with reduced effects on microtubule dynamic instability. *J Biol Chem,* 2005 280, 12902-12907.

[16] Luqmani, YA. Mechanisms of drug resistance in cancer chemotherapy. *Med Princ Pract,* 2005 14 Suppl 1, 35-48.

[17] Yusuf, RZ; Duan, Z; Lamendola, DE; Penson, RT; Seiden, MV. Paclitaxel resistance: molecular mechanisms and pharmacologic manipulation. *Curr Cancer Drug Targets,* 2003 3, 1-19.

[18] Kingston, DG. Tubulin-interactive natural products as anticancer agents. *J Nat Prod,* 2009 72, 507-15.

[19] Leistner, E. [Drugs from nature. The biology of taxane]. *Pharm Unserer Zeit, 2005* 34, 98-103.

[20] Seletsky, BM; Wang, Y; Hawkins, LD; Palme, MH; Habgood, GJ; DiPietro, LV; Towle, MJ; Salvato, KA; Wels, BF; Aalfs, KK; Kishi, Y; Littlefield, BA; Yu, MJ. Structurally simplified macrolactone analogues of halichondrin B. *Bioorg Med Chem Lett* 2004 14, 5547-5550.

[21] Mickel, SJ. Total synthesis of the marine natural product (+)-discodermolide in multigram quantities. *Pure Appl. Chem,* 2007 79, 685-700.

[22] Wood, KW; Cornwell, WD; Jackson, JR. Past and future of the mitotic spindle as an oncology target. *Curr Opin Pharmacol,* 2001 1,370-377.

[23] Shah, JV; Cleveland, DW. Waiting for anaphase: Mad2 and the spindle assembly checkpoint. *Cell,* 2000 103, 997-1000.

[24] Bharadwaj, R; Yu, H. The spindle checkpoint, aneuploidy, and cancer. *Oncogene,* 2004 23, 2016-2027.

[25] Kastan, MB; Bartek, J. Cell-cycle checkpoints and cancer. *Nature,* 2004 432, 316-323.

[26] Vermeulen, K; Van Bockstaele, DR; Berneman, ZN. The cell cycle: a review of regulation, deregulation and therapeutic targets in cancer. *Cell Prolif,* 2003 36:131-149.

[27] Santaguida, S; Musacchio, A. The life and miracles of kinetochores. This review serves as an excellent introduction to the structure and functions of the kinetochore. *EMBO J,* 2009 28, 2511-2531.

[28] Chan, GK; Liu, ST; Yen, TJ. Kinetochore structure and function. *Trends Cell Biol,* 2005 15, 589-598.

[29] Cheeseman, IM; Desai, A. Molecular architecture of the kinetochore-microtubule interface. *Nat Rev Mol Cell Biol,* 2008 9, 33-46.

[30] Clute, P; Pines, J. Temporal and spatial control of cyclin B1destruction in metaphase. *Nat Cell Biol,* 1999 1, 82-87.

[31] Peters, JM. The anaphase promoting complex/cyclosome: a machine designed to destroy. *Nat Rev Mol Cell Biol,* 2006 7, 644-656.

[32] Brito, DA; Rieder, CL. Mitotic checkpoint slippage in humans occurs via cyclin B destruction in the presence of an active checkpoint. *Curr Biol,* 2006 16, 1194-1200.

[33] Hengartner, MO. The biochemistry of apoptosis. *Nature,* 2000 407, 770-776.

[34] Huang, HC; Shi, J; Orth, JD; Mitchison, TJ. Cell death when the SAC is out of commission. *Cell Cycle,* 2010 9, 2049-2050.

[35] Wertz, IE; Kusam, S; Lam, C; Okamoto, T; Sandoval, W; Anderson, DJ; Helgason, E; Ernst, JA; Eby, M; Liu, J; Belmont, LD; Kaminker, JS; O'Rourke, KM; Pujara, K; Kohli, PB; Johnson, AR; Chiu, ML; Lill, JR; Jackson, PK; Fairbrother, WJ; Seshagiri, S; Ludlam, MJ; Leong, KG; Dueber, EC; Maecker, H; Huang, DC; Dixit, VM. Sensitivity to antitubulin chemotherapeutics is regulated by MCL1 and FBW7. *Nature* 2011 471, 110-114.

[36] Youle, RJ; Strasser, A. The BCL-2 protein family: opposing activities that mediate cell death. *Nat Rev Mol Cell Biol,* 2008 9, 47-59.

[37] Michels, J; Johnson, PW; Packham, G. Mcl-1. *Int J Biochem Cell Biol,* 2005 37, 267-271.

[38] Craig, RW. MCL1 provides a window on the role of the BCL2 family in cell proliferation, differentiation and tumorigenesis. *Leukemia,* 2002 16, 444-454.

[39] Schwickart, M; Huang, X; Lill, JR; Liu, J; Ferrando, R; French, DM; Maecker, H; O'Rourke, K; Bazan, F; Eastham-Anderson, J.

Deubiquitinase USP9X stabilizes MCL1 and promotes tumour cell survival. *Nature,* 2010 463, 103-107.

[40] Matson, DR; Stukenberg, PT. Spindle Poisons and Cell Fate: A Tale of Two Pathways. *Mol Interv,* 2011 2, 141-50.

[41] Suffness, M; Wall, ME. In *Taxol: Science and Applications*; Suffness, M., Eds.; CRC Press, Inc.: Boca Raton, FL, 1995; pp 3-25.

[42] Schiff, PB; Fant, J; Horwitz, SB. *Nature,* 1979 277, 665–667.

[43] Arbuck, SG; Blaylock, BA. In *Taxol: Science and Applications*; Suffness, M., Ed.; CRC Press, Inc.: Boca Raton, FL, 1995; pp 379-415.

[44] McGuire, W P; Rowinsky, EK; Rosenshein, NB; Grumbine FC; Ettinger, DS; Armstrong, DK; Donehower, RC. *Ann Intern Med,* 1989 111, 273–279.

[45] Holmes, FA; Walters, RS; Theriault, RL; Forman, AD; Newton, LK; Raber, MN; Buzdar, AU; Frye, DK; Hortobagyi, GN. *J Nat Cancer Inst,* 1991 83, 1797–1805.

[46] Ramalingam, S; Belani, C P. *Expert Opin Pharmacother* 2004 5, 1771–1780.

[47] Kavallaris, M; Kuo, DY; Burkhart, CA; Regl, DL; Norris, MD; Haber, M; Horwitz, SB. Taxol-resistant epithelial ovarian tumors are associated with altered expression of specific beta-tubulin isotypes. *J Clin Invest,* 1997 100, 1282-1293.

[48] Orr, G; Verdier-Pinard, P. Mechanisms of taxol resistance related to microtubules. *Oncogene,* 2003 22; 7280–7295.

[49] Honore, S; Pasquier, E; Braguer, D. Understanding microtubule dynamics for improved cancer therapy. *Cell Mol Life Sci,* 2005 62, 3039–3056.

[50] Cassimeris, L. The oncoprotein 18/stathmin family of microtubule destabilizers, *Curr Opin Cell Biol,* 2002 14, 18–24.

[51] Derry, W; Wilson, L; Khan, IA; Ludena, RF; Jordan, MA. Taxol differentially modulates the dynamics of microtubules assembled from unfractionated and purified beta-tubulin isotypes. *Biochemistry,* 1997 36, 3554–3562.

[52] Wang, TH; Wang, HS; Ichijo, H. Microtubule-interfering agents activate c-Jun-N-terminal kinase/stress-activated protein kinase through both Ras and apoptosis signal-regulating pathways. *J Biol Chem,* 1998 273, 4928–4936.

[53] Ibrado, AM; Kim, CN; Bhalla, K. Temporal relationship of CDK1 activation and mitotic arrest to cytosolic accumulation of cytochrome C

and caspase-3 activity during Taxol-induced apoptosis of human AML HL-60 cells. *Leukemia,* 1998 12, 1930–1936.

[54] Ling, YH; Consoli, U; Tornos, C; Andreeff, M; Perez-Soler, R. Accumulation of cyclin B1, activation of cyclin B1-dependent kinase and induction of programmed cell death in human epidermoid carcinoma KB cells treated with taxol. *Int J Cancer,* 1998 75, 925–32.

[55] Sudo, T; Nitta, M; Saya, H; Ueno, NT. Dependence of paclitaxel sensitivity on a functional spindle assembly checkpoint. *Cancer Res,* 2004 64, 2502–8.

[56] Wang, S; Wang, Z; Boise, L; Dent, P; Grant, S. Loss of the bcl-2 phosphorylation loop domain increases resistance of human leukemia cells (U937) to paclitaxel-mediated mitochondrial dysfunction and apoptosis. *Biochem Biophys Res Commun,* 1999 259, 67–72.

[57] Jordan, MA; Toso, RJ; Thrower, D; Wilson, L. Mechanism of mitotic block and inhibition of cell proliferation by taxol at low concentrations. *Proc Natl Acad Sci U S A,* 1993 90, 9552–9556.

[58] Long, BH; Fairchild, CR. Paclitaxel inhibits progression of mitotic cells to G1 phase by interference with spindle formation without affecting other microtubule functions during anaphase and telephase. *Cancer Res,* 1994 54, 4355–61.

[59] Wang, J; Lou, P; Lesniewski, R; Henkin, J. Paclitaxel at ultra low concentrations inhibits angiogenesis without affecting cellular microtubule assembly. *Anticancer Drugs,* 2003 14, 13–9.

[60] Wang, TH; Wang, HS; Soong, YK. Paclitaxel-induced cell death: where the cell cycle and apoptosis come together. *Cancer,* 2000 88, 2619–28.

[61] Weaver, BA; Cleveland, DW. Decoding the links between mitosis, cancer, and chemotherapy: the mitotic checkpoint, adaption, and cell death. *Cancer Cell,* 2005 8, 7–12.

[62] Rieder, CL; Maiato, H. Stuck in division or passing through: what happens when cells cannot satisfy the spindle assembly checkpoint. *Dev Cell,* 2004 7, 637–651.

[63] Blagosklonny, MV. Mitotic arrest and cell fate. *Cell Cycle,* 2007 6, 70–74.

[64] Thigpen, JT; Blessing, JA; Ball, H; Hummel, SJ; Barrett, RJ. Phase II trial of paclitaxel in patients with progressive ovarian carcinoma after platinum-based chemotherapy: a Gynecologic Oncology Group study. *J Clin Oncol,* 1994 12, 1748–1753.

[65] Einzig, AI. Review of phase II trials of Taxol (paclitaxel) in patients with advanced ovarian cancer. *Ann Oncol,* 1994 5, S29–32.

[66] Fountzilas, G; Athanassiades, A; Giannakakis, T; Bafaloukos, D; Karakousis, K; Dombros, N, Kosmidis, P; Skarlos, D. A phase II study of paclitaxel in advanced breast cancer resistant to anthracyclines. *Eur J Cancer,* 1996 32A, 47–51.

[67] Wilson, WH; Berg, SL; Bryant, G; Wittes, RE; Bates, S; Fojo, A; et al. Paclitaxel in doxorubicin-refractory or mitoxantrone-refractory breast cancer: a phase I/II trial of 96-hour infusion. *J Clin Oncol,* 1994 12, 1621–1629.

[68] Millward, MJ; Bishop, JF; Friedlander, M; Levi, JA; Goldstein, D; Olver, IN; Smith, JG; Toner, GC; Rischin, D; Bell, DR.Phase II trial of a 3-hour infusion of paclitaxel in previously untreated patients with advanced non-small-cell lung cancer. *J Clin Oncol,* 1996 14, 142–148.

[69] Gatzemeier, U; Heckmayr, M; Neuhauss, R; Schluter, I; Pawel, JV; Wagner, H; Dreps, A. Phase II study with paclitaxel for the treatment of advanced inoperable nonsmall cell lung cancer. *Lung cancer,* 1995 12, S101–106.

[70] Papamichael, D; Gallagher, CJ; Oliver, RT; Johnson, PW; Waxman, J. Phase II study of paclitaxel in pretreated patients with locally advanced/metastatic cancer of the bladder and ureter. *Br J Cancer,* 1997 75, 606–607.

[71] Murray, S; Briasoulis, E; Linardou, H; Bafaloukos, D; Papadimitriou, C. Taxane resistance in breast cancer: mechanisms, predictive biomarkers and circumvention strategies. *Cancer Treat Rev,* 2012 38, 890-903.

[72] Berrieman, HK; Lind, MJ; Cawkwell, L. Do beta-tubulin mutations have a role in resistance to chemotherapy? *Lancet Oncol,* 2004 5, 158-64.

[73] Wang, Y; O'Brate, A; Zhou, W; Giannakakou, P. Resistance to microtubule-stabilizing drugs involves two events, beta-tubulin mutation in one allele followed by loss of the second allele. *Cell Cycle,* 2005 4, 1847–1853.

[74] Martello, LA; Verdier-Pinard, P; Shen, HJ; He, L; Torres, K; Orr, GA; Horwitz, SB. Elevated levels of microtubule destabilizing factors in a taxol resistant/dependent A549 cell line with an {alpha}-tubulin mutation. *Cancer Res,* 2003 63, 1207–1213.

[75] Wagner, P; Wang, B; Clark, E; Lee, H; Rouzier, R; Pusztai, L. Microtubule associated protein (MAP)-tau a novel mediator of paclitaxel in vitro and in vivo. *Cell Cycle,* 2005 4, 1149–1152.

[76] Rouzier, R; Rajan, R; Wagner, P; Hess, KR; Gold, DL; Stec, J; Ayers, M; Ross, JS; Zhang, P; Buchholz, TA; Kuerer, H; Bgreen, M; Arun, B;

Hortobagyi, GN; Symmans, WF; Pusztai, L. Microtubule associated protein tau: a marker of paclitaxel sensitivity in breast cancer. *Proc Natl Acad Sci USA,* 2005 102, 8315–8320.

[77] Hawkins, D; Demers, G; Galloway, DA. Inactivation of p53 enhances sensitivity to multiple chemotherapeutic agents. *Cancer Res,* 1996 56, 892–898.

[78] Zhang, CC; Yang, JM; White, E; Murphy, M; Levine, A; Hait, WN. The role of MAP4 expression in the sensitivity to paclitaxel and resistance to vinca alkaloids in p53 mutant cells. *Oncogene,* 1998 16, 1617–1624.

[79] Strobel, T; Kraeft, SK; Chen, LB; Cannistra, SA. BAX expression is associated with enhanced intracellular accumulation of paclitaxel: a novel role for BAX during chemotherapy-induced cell death. *Cancer Res,* 1998 58, 4776–4781.

[80] Giannakakou, P; Poy, G; Zhan, Z; Knutsen, T; Blagosklonny, MV; Fojo, T. Paclitaxel selects for mutant or pseudo-null p53 in drug resistance associated with tubulin mutations in human cancer. *Oncogene,* 2000 19, 3078–3085.

[81] Lanni, JS; Lowe, SW; Licitra, EJ; Liu, JO; Jacks, T. p53-independent apoptosis induced by paclitaxel through an indirect mechanism. *Proc Natl Acad Sci USA,* 1997 94, 9679–9683.

[82] Huang, Y; Ibrado, AM; Reed, JC; Bullock, G; Ray, S; Tang, C; Bhalla, K. Co-expression of several molecular mechanisms of multidrug resistance and their significance for paclitaxel cytotoxicity in human AML HL-60 cells. *Leukemia,* 1997 11, 253–257.

[83] Strobel, T; Swanson, L; Korsmeyer, S; Cannistra, SA. BAX enhances paclitaxelinduced apoptosis through a p53-independent pathway. *Proc Natl Acad Sci USA,* 1996 93, 14094–14099.

[84] Noguchi, S. Predictive factors for response to docetaxel in human breast cancers. *Cancer Sci,* 2006 97, 813–20.

[85] Wang, Z; Goulet 3rd, R; Stanton, KJ; Sadaria, M; Nakshatri, H. Differential effect of anti-apoptotic genes Bcl-xL and c-FLIP on sensitivity of MCF-7 breast cancer cells to paclitaxel and docetaxel. *Anticancer Res,* 2005 25, 2367–2379.

[86] Ganta, S; Amiji, M. Coadministration of Paclitaxel and curcumin in nanoemulsion formulations to overcome multidrug resistance in tumor cells. *Mol Pharm,* 2009 6, 928–39.

[87] Mabuchi, S; Ohmichi, M; Nishio, Y; Hayasaka, T; Kimura, A; Ohta, T; Kawagoe, J; Takahashi, K; Yada-Hashimoto, N; Seino-Noda, H; Sakata, M; Motoyama, T; Kurachi, H; Testa, JR; Tasaka, K; Murata, Y.

Inhibition of inhibitor of nuclear factor-kappaB phosphorylation increases the efficacy of paclitaxel in in vitro and in vivo ovarian cancer models. *Clin Cancer Res, 2004* 10, 7645–7654.

[88] Knuefermann, C; Lu, Y; Liu, B; Jin, W; Liang, K; Wu, L; Schmidt, M; Mills, GB; Mendelsohn, J; Fan, Z. HER2/PI-3K/Akt activation leads to a multidrug resistance in human breast adenocarcinoma cells. *Oncogene,* 2003 22, 3205–3212.

[89] Sastre-Garau, X; Genin, P; Rousseau, A; Al Ghuzlan, A; Nicolas, A; Freneaux, P; Rosty, C; Sigal-Zafrani, B; Couturier, J; Thiery, JP; Magdelénat, H; Vincent-Salomon, A. Increased cell size and Akt activation in HER-2/neu-overexpressing invasive ductal carcinoma of the breast. *Histopathology,* 2004 45, 142–147.

[90] Skoufias, DA; Andreassen, PR; Lacroix, FB; Wilson, L; Margolis, RL. Mammalian mad2 and bub1/bubR1 recognize distinct spindle-attachment and kinetochore-tension checkpoints. *Proc Natl Acad Sci USA,* 2001 98, 4492–4497.

[91] Takahashi, T; Yamasaki, F; Sudo, T; Itamochi, H; Adachi, S; Tamamori-Adachi, M, Ueno, NT. Cyclin A-associated kinase activity is needed for paclitaxel sensitivity. *Mol Cancer Ther,* 2005 4, 1039–1046.

[92] Chabalier, C; Lamare, C; Racca, C; Privat, M; Valette, A; Larminat, F. BRCA1 downregulation leads to premature inactivation of spindle checkpoint and confers paclitaxel resistance. *Cell Cycle,* 2006 5, 1001–1007.

[93] Alli, E; Yang, JM; Ford, JM; Hait, WN. Reversal of stathmin-mediated resistance to paclitaxel and vinblastine in human breast carcinoma cells. *Mol Pharmacol,* 2007 71, 1233–1240.

[94] Chuthapisith, S; Layfield, R; Kerr, ID; Hughes, C; Eremin, O. Proteomic profiling of MCF-7 breast cancer cells with chemoresistance to different types of anticancer drugs. *Int J Oncol,* 2007 30, 1545–1551.

[95] Yu, D; Jing, T; Liu, B; Yao, J; Tan, M; McDonnell, TJ; Hung, MC. Overexpression of ErbB2 blocks Taxol-induced apoptosis by upregulation of p21Cip1, which inhibits p34Cdc2 kinase. *Mol Cell,* 1998 2, 581–591.

[96] Tan, M; Jing, T; Lan, KH; Neal, CL; Li, P; Lee, S; Fang, D; Nagata, Y; Liu, J; Arlinghaus, R; Hung, MC; Yu, D. Phosphorylation on tyrosine-15 of p34(Cdc2) by ErbB2 inhibits p34(Cdc2) activation and is involved in resistance to taxol-induced apoptosis. *Mol Cell,* 2002 9, 993–1004.

[97] Slamon, DJ; Clark, GM; Wong, SG; Levin, WJ; Ullrich, A; McGuire, WL. Human breast cancer: correlation of relapse and survival with amplification of the HER-2/neu oncogene. *Science,* 1987 235, 177–182.

[98] Nagai, MA; Marques, LA; Torloni, H; Brentani, MM. Genetic alterations in cerbB-2 protooncogene as prognostic markers in human primary breast tumors. *Oncology,* 1993 50, 412–417.

[99] de Munck, L; Schaapveld, M; Siesling, S; Wesseling, J; Voogd, AC; Tjan-Heijnen, VC; Otter, R; Willemse, PH. Implementation of trastuzumab in conjunction with adjuvant chemotherapy in the treatment of non-metastatic breast cancer in the Netherlands. *Breast Cancer Res Treat,* 2010 129, 229–233.

[100] Azambuja, E; Durbecq, V; Rosa, DD; Colozza, M; Larsimont, D; Piccart-Gebhart, M; Cardoso, F. HER-2 overexpression/amplification and its interaction with taxanebased therapy in breast cancer. *Ann Oncol,* 2008 19, 223–232.

[101] Ciardiello, F; Caputo, R; Pomatico, G; De Laurentiis, M; De Placido, S; Bianco, AR; Tortora, G. Resistance to taxanes is induced by c-erbB-2 overexpression in human MCF-10A mammary epithelial cells and is blocked by combined treatment with an antisense oligonucleotide targeting type I protein kinase A. *Int J Cancer,* 2000 85, 710–715.

[102] Yu, D; Liu, B; Jing, T; Sun, D; Price, JE; Singletary, SE; Ibrahim, N;. Hortobagyi, GN; Hung, MC. Overexpression of both p185c-erbB2 and p170mdr-1 renders breast cancer cells highly resistant to taxol. *Oncogene,* 1998 16, 2087–2094.

[103] O'Brien, C; Cavet, G; Pandita, A; Hu, X; Haydu, L; Mohan, S; Toy, K; Rivers, CS; Modrusan, Z; Amler, LC; Lackner, MR. Functional genomics identifies ABCC3 as a mediator of taxane resistance in HER2-amplified breast cancer. *Cancer Res,* 2008 68, 5380–5389.

[104] Ling, V. P-glycoprotein and resistance to anticancer drugs. *Cancer,* 1992 69, 2603–2609.

[105] Schinkel, AH. The physiological function of drug-transporting P-glycoproteins. *Semin Cancer Biol,* 1997 8, 161–170.

[106] Cordon-Cardo, C; O'Brien, JP; Boccia, J; Casals, D; Bertino, JR; Melamed, MR. Expression of the multidrug resistance gene product (P-glycoprotein) in human normal and tumor tissues. *J Histochem Cytochem,* 1990 38, 1277–1287.

[107] Trock, BJ; Leonessa, F; Clarke, R. Multidrug resistance in breast cancer: a metaanalysis of MDR1/gp170 expression and its possible functional significance. *J Natl Cancer Inst,* 1997 89, 917–931.

[108] Dean, M; Rzhetsky, A; Allikmets, R. The human ATP-binding cassette (ABC) transporter superfamily. *Genome Res,* 2001 11, 1156–1166.
[109] Fojo, AT; Menefee, M. Microtubule targeting agents: basic mechanisms of multidrug resistance (MDR). *Semin Oncol,* 2005 32, S3–8.
[110] Szakacs, G; Annereau, JP; Lababidi, S; Shankavaram, U; Arciello, A; Bussey, KJ; Reinhold, W; Guo, Y; Kruh, GD; Reimers, M; Weinstein, JN; Gottesman, MM. Predicting drug sensitivity and resistance. profiling ABC transporter genes in cancer cells. *Cancer cell,* 2004 6, 129–137.
[111] Kruh, GD; Belinsky, MG. The MRP family of drug efflux pumps. *Oncogene,* 2003 22, 7537–7552.
[112] Gandhi, YA; Morris, ME. Structure-activity relationships and quantitative structure-activity relationships for breast cancer resistance protein (ABCG2). *Aaps J,* 2009 11, 541–552.
[113] Bell, DR; Trent, JM; Willard, HF; Riordan, JR; Ling, V. Chromosomal location of human P-glycoprotein gene sequences. *Cancer Genet Cytogenet,* 1987 25, 141–148.
[114] Kallio, MJ; Nieminen, M; Eriksson, JE. Human inhibitor of apoptosis protein (IAP) survivin participates in regulation of chromosome segregation and mitotic exit. *FASEB J,* 2001 15, 2721–2723.
[115] Li, F; Ambrosini, G; Chu, EY; Plescia, J; Tognin, S; Marchisio, PC; Altieri, DC. Control of apoptosis and mitotic spindle checkpoint by survivin. *Nature,* 1998 396, 580–584.
[116] Asanuma, H; Torigoe, T; Kamiguchi, K; Hirohashi, Y; Ohmura, T; Hirata, K; Sato, M; Sato, N. Survivin expression is regulated by coexpression of human epidermal growth factor receptor 2 and epidermal growth factor receptor via phosphatidylinositol 3-kinase/AKT signaling pathway in breast cancer cells. *Cancer Res,* 2005 65, 11018–11025.
[117] Ryan, BM; Konecny, GE; Kahlert, S; Wang, HJ; Untch, M; Meng, G; Pegram, MD; Podratz, KC; Crown, J; Slamon, DJ; Duffy, MJ. Survivin expression in breast cancer predicts clinical outcome and is associated with HER2, VEGF, urokinase plasminogen activator and PAI-1. *Ann Oncol* 2006 17, 597–604.
[118] Xia, W; Bisi, J; Strum, J; Liu, L; Carrick, K; Graham KM; Treece, AL; Hardwicke, MA; Dush, M; Liao, Q; Westlund, RE; Zhao, S; Bacus, S; Spector, NL. Regulation of survivin by ErbB2 signaling: therapeutic implications for ErbB2-overexpressing breast cancers. *Cancer Res,* 200) 66, 1640–1647.

[119] Bieche, I; Onody, P; Tozlu, S; Driouch, K; Vidaud, M; Lidereau, R. Prognostic value of ERBB family mRNA expression in breast carcinomas. *Int J Cancer,* 200) 106, 758–765.

[120] deFazio, A; Chiew, YE; Sini, RL; Janes, PW; Sutherland, RL. Expression of c-erbB receptors, heregulin and oestrogen receptor in human breast cell lines. *Int J Cancer,* 2000 87, 487–498.

[121] Clark, AS; West, K; Streicher, S; Dennis, PA. Constitutive and inducible Akt activity promotes resistance to chemotherapy, trastuzumab, or tamoxifen in breast cancer cells. *Mol Cancer Ther,* 2002 1, 707–717.

[122] Priulla, M; Calastretti, A; Bruno, P; Azzariti, A; Paradiso, A; Canti, G; Nicolin, A. Preferential chemosensitization of PTEN-mutated prostate cells by silencing the Akt kinase. *Prostate,* 2007 67, 782-789.

[123] Knuefermann, C; Lu, Y; Liu, B; Jin, W; Liang, K; Wu, L; Schmidt, M; Mills, GB; Mendelsohn, J; Fan, Z. HER2/PI-3K/Akt activation leads to a multidrug resistance in human breast adenocarcinoma cells. *Oncogene,* 2003 22, 3205–3212.

[124] Wang, S; Huang, X; Lee, CK; Liu, B. Elevated expression of erbB3 confers paclitaxel resistance in erbB2-overexpressing breast cancer cells via upregulation of Survivin. *Oncogene,* 2010 29, 4225–4236.

[125] Janz, M; Harbeck, N; Dettmar, P; Berger, U; Schmidt, A; Jurchott, K; Schmitt, M; Royer, HD. Y-box factor YB-1 predicts drug resistance and patient outcome in breast cancer independent of clinically relevant tumor biologic factors HER2, uPA and PAI-1. *Int J Cancer,* 2002 97, 278–282.

[126] Bargou, RC; Jurchott, K; Wagener, C; Bergmann, S; Metzner, S; Bommert, K; Mapara, MY; Winzer, KJ; Dietel, M; Dorken, B; Royer, HD. Nuclear localization and increased levels of transcription factor YB-1 in primary human breast cancers are associated with intrinsic MDR1 gene expression. *Nat Med,* 1997 3, 447–450.

[127] Rubinstein, DB; Stortchevoi, A; Boosalis, M; Ashfaq, R; Guillaume, T. Overexpression of DNA-binding protein B gene product in breast cancer as detected by in vitro-generated combinatorial human immunoglobulin libraries. *Cancer Res,* 2002 62, 4985–4991.

[128] Sutherland, BW; Kucab, J; Wu, J; Lee, C; Cheang, MC; Yorida, E; Turbin, D; Dedhar, S; Nelson, C; Pollak, M; Leighton Grimes, H; Miller, K; Badve, S; Huntsman, D; Blake-Gilks, C; Chen, M; Pallen, CJ; Dunn, SE. Akt phosphorylates the Y-box binding protein 1 at Ser102 located in the cold shock domain and affects the anchorage-

independent growth of breast cancer cells. *Oncogene,* 2005 24, 4281–4292.

[129] Wu, J; Lee, C; Yokom, D; Jiang, H; Cheang, MC; Yorida, E; Turbin, D; Berquin, IM; Mertens, PR; Iftner, T; Gilks, CB; Dunn, SE. Disruption of the Y-box binding protein-1 results in suppression of the epidermal growth factor receptor and HER-2. *Cancer Res,* 2006 66, 4872–4879.

[130] Chatterjee, M; Rancso, C; Stuhmer, T; Eckstein, N; Andrulis, M; Gerecke, C; Lorentz, H; Royer, HD; Bargou, RC. The Y-box binding protein YB-1 is associated with progressive disease and mediates survival and drug resistance in multiple myeloma. *Blood,* 2008 111, 3714–3722.

[131] Fujita, T; Ito, K; Izumi, H; Kimura, M; Sano, M; Nakagomi, H; Maeno, K; Hama, Y; Shingu, K; Tsuchiya, S; Kohno, K; Fujimori, M. Increased nuclear localization of transcription factor Y-box binding protein 1 accompanied by up-regulation of P-glycoprotein in breast cancer pretreated with paclitaxel. *Clin Cancer Res,* 2005 11, 8837–8844.

[132] Lee, C; Dhillon, J; Wang, MY; Gao, Y; Hu, K; Park, E; Astanehe, A; Hung, MC; Eirew, P; Eaves, CJ; Dunn, SE. Targeting YB-1 in HER-2 overexpressing breast cancer cells induces apoptosis via the mTOR/STAT3 pathway and suppresses tumor growth in mice. *Cancer Res,* 2008 68, 8661–8666.

[133] Coles, BF; Kadlubar, FF. Detoxification of electrophilic compounds by glutathione S-transferase catalysis: determinants of individual response to chemical carcinogens and chemotherapeutic drugs? *Biofactors,* 2003 17, 115–130.

[134] Huang, J; Tan, PH; Thiyagarajan, J; Bay, BH. Prognostic significance of glutathione S-transferase-pi in invasive breast cancer. *Mod Pathol,* 2003 16, 558–565.

[135] Silvestrini, R; Veneroni, S; Benini, E; Daidone, MG; Luisi, A; Leutner, M; Maucione, A; Kenda, R; Zucali, R; Veronesi, U. Expression of p53, glutathione Stransferase-pi, and Bcl-2 proteins and benefit from adjuvant radiotherapy inbreast cancer. *J Natl Cancer Inst,* 1997 89, 639–645.

[136] Su, F; Hu, X; Jia, W; Gong, C; Song, E; Hamar, P. Glutathion S transferase pi indicates chemotherapy resistance in breast cancer. *J Surg Res* 2003 113, 102–108.

[137] Iwao-Koizumi, K; Matoba, R; Ueno, N; Kim, SJ; Ando, A; Miyoshi, Y; Maeda, E; Noguchi, S; Kato, K. Prediction of docetaxel response in

human breast cancer by gene expression profiling. *J Clin Oncol,* 2005 23, 422–431.

[138] Miyoshi, Y; Taguchi, T; Kim, SJ; Tamaki, Y; Noguchi, S. Prediction of response to docetaxel by immunohistochemical analysis of CYP3A4 expression in human breast cancers. *Breast Cancer,* 2005 12, 11–15.

[139] Miyoshi, Y; Ando, A; Takamura, Y; Taguchi, T; Tamaki, Y; Noguchi, S. Prediction of response to docetaxel by CYP3A4 mRNA expression in breast cancer tissues. *Int J Cancer* 2002 97, 129–132.

[140] McDonald, SL; Stevenson, DA; Moir, SE; Hutcheon AW; Haites, NE; Heys, SD; Schofield, AC. Genomic changes identified by comparative genomic hybridisation in docetaxel-resistant breast cancer cell lines. *Eur J Cancer* 2005 41, 1086–1094.

[141] Amirimani, B; Ning, B; Deitz, AC; Weber, BL; Kadlubar, FF; Rebbeck, TR. Increased transcriptional activity of the CYP3A4*1B promoter variant. *Environ Mol Mutagen* 2003 42, 299–305.

[142] Schirmer, M; Toliat, MR; Haberl, M; Suk, A; Kamdem, LK; Klein, K; Brockmoller, J; Nurnberg, P; Zanger, UM; Wojnowski, L. Genetic signature consistent with selection against the CYP3A4*1B allele in non-African populations, *Pharmacogenet. Genomics,* 2006 16, 59–71.

[143] Tran, A; Jullien, V; Alexandre, J; Rey, E; Rabillon, F; Girre, V; Dieras, V; Pons, G; Goldwasser, F; Treluyer, JM. Pharmacokinetics and toxicity of docetaxel: role of CYP3A, MDR1, and GST polymorphisms. *Clin Pharmacol Ther* 2006 79, 570–580.

[144] Inuzuka, H; Shaik, S; Onoyama, I; Gao, D; Tseng, A; Maser, RS; Zhai, B; Wan, L; Gutierrez, A; Lau, AW; Xiao, Y; Christie, AL; Aster, J; Settleman, J; Gygi, SP; Kung, AL; Look, T; Nakayama, KI; DePinho RA, Wei W. SCF (FBW7) regulates cellular apoptosis by targeting MCL1 for ubiquitylation and destruction. *Nature,* 2011 471, 104–109.

[145] Hu, L; Hofmann, J; Lu, Y; Mills, GB; Jaffe, RB. Inhibition of phosphatidylinositol 3'-kinase increases efficacy of paclitaxel in in vitro and in vivo ovarian cancer models. *Cancer Res,* 2002 62, 1087-1092.

[146] Mackeigan, JP; Collins, TS; Ting, JP-Y. MEK inhibition enhances paclitaxel-induced tumor apoptosis. *J Biol Chem* 2000 275, 38953–38956.

[147] Chang, JC; Wooten, EC; Tsimelzon, A; Hilsenbeck, SG; Gutierrez, MC; Elledge, R; Mohsin, S; Osborne, CK; Chamness, GC; Allred, DC; O'Connell, P. Gene expression profiling for the prediction of therapeutic response to docetaxel in patients with breast cancer. *Lancet,* 2003 362, 362–369.

[148] Yared, JA; Tkaczuk, KH. Update on taxane development: new analogs and new formulations. *Drug Des Devel Ther,* 2012 6, 371-384.

[149] ten Tije, AJ; Verweij, J; Loos, WJ; Sparreboom, A. Pharmacological effects of formulation vehicles : implications for cancer chemotherapy. *Clin Pharmacokinet* 2003 42, 665-85.

[150] Ibrahim, NK; Desai, N; Legha, S; Soon-Shiong, P; Theriault, RL; Rivera, E; Esmaeli, B; Ring, SE; Bedikian, A; Hortobagyi, GN; Ellerhorst, JA. Phase I and pharmacokinetic study of ABI-007, a Cremophor-free, protein-stabilized, nanoparticle formulation of paclitaxel. *Clin Cancer Res,* 2002 8, 1038-44.

[151] Desai, N; Trieu, V; Yao, Z; Louie, L; Ci, S; Yang, A; Tao, C; De, T; Beals, B; Dykes, D; Noker, P; Yao, R; Labao, E; Hawkins, M; Soon-Shiong, P. Increased antitumor activity, intratumor paclitaxel concentrations, and endothelial cell transport of cremophor-free, albumin-bound paclitaxel, ABI-007, compared with cremophor-based paclitaxel. *Clin Cancer Res,* 2006 12, 1317-1324.

[152] Hawkins, MJ; Soon-Shiong, P; Desai, N. Protein nanoparticles as drug carriers in clinical medicine. *Adv Drug Deliv Rev,* 2008 60, 876-885.

[153] Pal, SK; Twardowski, P; Sartor, O. Critical appraisal of cabazitaxel in the management of advanced prostate cancer. *Clin Interv Aging,* 2010 5, 395-402.

[154] Shionoya, M; Jimbo, T; Kitagawa, M; Soga, T; Tohgo, A. DJ-927, a novel oral taxane, overcomes P-glycoprotein-mediated multidrug resistance in vitro and in vivo. *Cancer Sci,* 2003 94, 459-466.

[155] Singer, JW; Baker, B; De Vries, P; Kumar, A; Shaffer, S; Vawter, E; Bolton, M; Garzone, P. Poly-(L)-glutamic acid-paclitaxel (CT-2103) [XYOTAX], a biodegradable polymeric drug conjugate: characterization, preclinical pharmacology, and preliminary clinical data. *Adv Exp Med Biol,* 2003 519, 81-99.

[156] Geney, R; Chen, J; Ojima, I. Recent advances in the new generation taxane anticancer agents. *Med Chem,* 2005 1, 125-139.

[157] Bissery, MC. Preclinical evaluation of new taxoids. *Curr Pharm* Des, 2001 7, 1251–1257.

[158] Kim, SC; Kim, DW; Shim, YH; Bang, JS; Oh, HS; Wan Kim, S; Seo, MH. In vivo evaluation of polymeric micellar paclitaxel formulation: toxicity and efficacy. *J Control Release* 2001 72, 191-202.

[159] Negishi, T; Koizumi, F; Uchino, H; Kuroda, J; Kawaguchi, T; Naito, S; Matsumura, Y. NK105, a paclitaxel-incorporating micellar nanoparticle,

is a more potent radiosensitising agent compared to free paclitaxel. *Br J Cancer,* 2006 95, 601-606.

[160] Bradley, MO; Swindell, CS; Anthony, FH; Witman, PA; Devanesan, P; Webb, NL; Baker, SD; Wolff, AC; Donehower, RC. Tumor targeting by conjugation of DHA to paclitaxel. *J Control Release,* 2001 74, 233-236.

[161] Régina, A; Demeule, M; Ché, C; Lavallée, I; Poirier, J; Gabathuler, R; Béliveau, R; Castaigne, JP. Antitumour activity of ANG1005, a conjugate between paclitaxel and the new brain delivery vector Angiopep-2. *Br J Pharmacol,* 2008 155, 185-197.

[162] Kingston, DG; Newman, DJ. Taxoids: cancer-fighting compounds from nature. *Curr Opin Drug Discov Devel,* 2007 10, 130-144.

[163] Nobili, S; Landini, I; Mazzei, T; Mini, E. Overcoming tumor multidrug resistance using drugs able to evade P-glycoprotein or to exploit its expression. *Med Res Rev,* 2012 32, 1220-1262.

[164] Teodori, E; Dei, S; Scapecchi, S; Gualtieri, F. Corresponding author contact information. The medicinal chemistry of multidrug resistance (MDR) reversing drugs. *Il Farmaco,* 2002 57, 385–415.

[165] Tsuruo, T; Iida, H; Tsukagoshi, S; Sakurai, Y. Overcoming of vincristine resistance in P388 leukemia in vivo and in vitro through enhanced cytotoxicity of vincristine and vinblastine by verapamil. *Cancer Res,* 41, 1967-1972.

[166] Tarasova, NI; Seth, R; Tarasov, SG; Kosakowska-Cholody, T; Hrycyna, CA; Gottesman, MM; Michejda, CJ. Transmembrane inhibitors of P-glycoprotein, an ABC transporter. *J Med Chem,* 2005 48, 3768-3775.

[167] Kang, H; Fisher, MH; Xu, D; Miyamoto, YJ; Marchand, A; Van Aerschot, A; Herdewijn, P; Juliano, RL. Inhibition of MDR1 gene expression by chimeric HNA antisense oligonucleotides. *Nucleic Acids Res,* 2004 32, 4411-4419.

[168] Takara, K; Sakaeda, T; Okumura, K. An update on overcoming MDR1-mediated multidrug resistance in cancer chemotherapy. *Curr Pharm Des,* 2006 12, 273-286.

[169] Dačević, M; Isaković, A; Podolski-Renić, A; Isaković, AM; Stanković, T; Milošević, Z; Rakić, L; Ruždijić, S; Pešić, M. Purine nucleoside analog--sulfinosine modulates diverse mechanisms of cancer progression in multi-drug resistant cancer cell lines. *PLoS One,* 2013 8, e54044.

[170] Blagosklonny, MV. Why therapeutic response may not prolong the life of a cancer patient: selection for oncogenic resistance. *Cell Cycle,* 2005 4, 1693-1698.

[191] Rose, WC; Long, BH; Fairchild, CR; Lee, FY; Kadow, JF. Preclinical pharmacology of BMS-275183, an orally active taxane. *Clin Cancer Res,* 2001 7, 2016-2021.
[192] Giannakakou, P; Sackett, DL; Kang, YK; Zhan, Z; Buters, JT; Fojo, T; Poruchynsky, MS. Paclitaxel-resistant human ovarian cancer cells have mutant beta-tubulins that exhibit impaired paclitaxel-driven polymerization. *J Biol Chem,* 1997 272, 17118-17125.
[193] Sampath, D; Discafani, CM; Loganzo, F; Beyer, C; Liu, H; Tan, X; Musto, S; Annable, T; Gallagher, P; Rios, C; Greenberger, LM. MAC-321, a novel taxane with greater efficacy than paclitaxel and docetaxel in vitro and in vivo. *Mol Cancer Ther,* 2003 2, 873-884.
[194] Hari, M; Loganzo, F; Annable, T; Tan, X; Musto, S; Morilla, DB; Nettles, JH; Snyder, JP; Greenberger, LM. Paclitaxel-resistant cells have a mutation in the paclitaxel-binding region of beta-tubulin (Asp26Glu) and less stable microtubules. *Mol Cancer Ther,* 2006 5, 270-278.
[195] Sampath, D; Greenberger, LM; Beyer, C; Hari, M; Liu, H; Baxter, M; Yang, S; Rios, C; Discafani, C. Preclinical pharmacologic evaluation of MST-997, an orally active taxane with superior in vitro and in vivo efficacy in paclitaxel- and docetaxel-resistant tumor models. *Clin Cancer Res,* 2006 12, 3459-3469.
[196] Ferlini, C; Raspaglio, G; Mozzetti, S; Cicchillitti, L; Filippetti, F; Gallo, D; Fattorusso, C; Campiani, G; Scambia, G. The seco-taxane IDN5390 is able to target class III beta-tubulin and to overcome paclitaxel resistance. *Cancer Res,* 2005 65, 2397-2405.
[197] Dumontet, C; Jordan, MA. Microtubule-binding agents: a dynamic field of cancer therapeutics. *Nature Reviews Drug Discovery,* 2009 9, 790-803.
[198] Shapira, A; Livney, YD; Broxterman, HJ; Assaraf, YG. Nanomedicine for targeted cancer therapy: towards the overcoming of drug resistance. *Drug Resist Updat,* 2011 14, 150-163.
[199] Wang, Z; Li, Y; Ahmad, A; Banerjee, S; Azmi, AS; Kong, D; Sarkar, FH. Pancreatic cancer: understanding and overcoming chemoresistance. *Nat Rev Gastroenterol Hepatol,* 2011 8, 27-33.
[200] Fletcher, JI; Haber, M; Henderson, MJ; Norris, MD. ABC transporters in cancer: more than just drug efflux pumps. *Nat Rev Cancer,* 2010 10, 147-156.
[201] Gottesman, MM; Fojo, T; Bates, SE. Multidrug resistance in cancer: role of ATP-dependent transporters. *Nat Rev Cancer,* 2002 2, 48-58.

[202] Burger, H; Loos, WJ; Eechoute, K; Verweij, J; Mathijssen, RH; Wiemer, EA. Drug transporters of platinum-based anticancer agents and their clinical significance. *Drug Resist Updat,* 2011 14, 22-34.
[203] Singh, A; Settleman, J. EMT, cancer stem cells and drug resistance: an emerging axis of evil in the war on cancer. *Oncogene,* 2010 29, 4741-4751.
[204] Marin, JJ; Castaño, B; Blazquez, AG; Rosales, R; Efferth, T; Monte, MJ. Strategies for overcoming chemotherapy resistance in enterohepatic tumours. *Curr Mol Med,* 2010 10, 467-485.
[205] Dean, M; Allikmets, R. Complete characterization of the human ABC gene family. *J Bioenerg Biomembr,* 2001 33, 475-479.
[206] Ding, XW; Wu, JH; Jiang, CP. ABCG2: a potential marker of stem cells and novel target in stem cell and cancer therapy. *Life Sci,* 2010 86, 631-637.
[207] Valero, V; Hortobagyi, GN. Primary chemotherapy: a better overall therapeutic option for patients with breast cancer. *Ann Oncol,* 1998 9, 1151-1154.
[208] Szakács, G; Paterson, JK; Ludwig, JA; Booth-Genthe, C; Gottesman, MM. Targeting multidrug resistance in cancer. *Nat Rev Drug Discov,* 2006 5, 219-234.
[209] Trock, BJ; Leonessa, F; Clarke, R. Multidrug resistance in breast cancer: a meta-analysis of MDR1/gp170 expression and its possible functional significance. *J Natl Cancer Inst,* 1997 89, 917-931.
[210] Soussi, T; Rubio-Nevado, JM; Ishioka, C. MUT-TP53: a versatile matrix for TP53 mutation verification and publication. *Hum Mutat,* 2006 27, 1151-1154.
[211] Vincent-Salomon, A; Rousseau, A; Jouve, M; Beuzeboc, P; Sigal-Zafrani, B; Fréneaux, P; Rosty, C; Nos, C; Campana, F; Klijanienko, J; Al Ghuzlan, A; Sastre-Garau, X. Proliferation markers predictive of the pathological response and disease outcome of patients with breast carcinomas treated by anthracycline-based preoperative chemotherapy. *Eur J Cancer,* 2004 40, 1502-1508.
[212] Daidone, MG; Silvestrini, R. Prognostic and predictive role of proliferation indices in adjuvant therapy of breast cancer. *J Natl Cancer Inst Monogr,* 2001 30, 27-35.
[213] Amadori, D; Volpi, A; Maltoni, R; Nanni, O; Amaducci, L; Amadori, A; Giunchi, DC; Vio, A; Saragoni, A; Silvestrini, R. Cell proliferation as a predictor of response to chemotherapy in metastatic breast cancer: a prospective study. *Breast Cancer Res Treat,* 1997 43, 7-14.

[214] Makin, G; Dive, C. Modulating sensitivity to drug-induced apoptosis: the future for chemotherapy? *Breast Cancer Res* 2001 3, 150-153.

[215] Rigas, RJ. Taxane-Platinum Combinations in Advanced Non-Small Cell Lung Cancer: A Review. *The Oncologist,* 2004 9, 16-23.

[216] Kosmidis, P; Mylonakis, N; Nicolaides, C; Kalophonos, C; Samantas, E; Boukovinas, J; Fountzilas, G; Skarlos, D; Economopoulos, T; Tsavdaridis, D; Papakostas, P; Bacoyiannis, C; Dimopoulos, M. Paclitaxel plus carboplatin versus gemcitabine plus paclitaxel in advanced non-small-cell lung cancer: a phase III randomized trial. *J Clin Oncol,* 2002 20, 3578-3585.

[217] Athanasiadis, A; Roussos, G; Papakostoulis, T; Athanasiadou, DC; Stathopoulos, GP. Paclitaxel and gemcitabine combination in a biweekly schedule in patients with advanced non small-cell lung cancer: a phase I study. *Cancer Chemother Pharmacol,* 2005 56, 653-658.

[218] Nabholtz JM. Docetaxel-anthracycline combinations in metastatic breast cancer. *Breast Cancer Res Treat* 2003 79, S3-9.

[219] Ho, YP; Au-Yeung, SC; To, KK. Platinum-based anticancer agents: innovative design strategies and biological perspectives. *Med Res Rev,* 2003 23, 633-655.

[220] Manegold, C; Zatloukal, P; Krejcy, K; Blatter, J. Gemcitabine in non-small cell lung cancer (NSCLC). *Invest New Drugs,* 2000 18, 29-42.

[221] Gewirtz, DA. A critical evaluation of the mechanisms of action proposed for the antitumor effects of the anthracycline antibiotics adriamycin and daunorubicin. *Biochem Pharmacol,* 1999 57, 727-741.

[222] Paterson, I; Anderson, EA. The renaissance of natural products as drug candidates. *Science,* 2005 310, 451-453.

[223] Cragg, GM; Grothaus, PG; Newman, DJ. Impact of natural products on developing new anti-cancer agents. *Chem Rev,* 2009 109, 3012-3043.

[224] Wartmann, M; Altmann, KH. The biology and medicinal chemistry of epothilones. *Curr Med Chem Anticancer Agents* 2002 2, 123-148.

[225] Gerth, K; Washausen, P; Hofle, G; Irschik, H; Reichenbach, H. The jerangolids: A family of new antifungal compounds from Sorangium cellulosum (Myxobacteria). Production, physico-chemical and biological properties of jerangolid A. *J Antibiot* (Tokyo), 1996 49, 71-75.

[226] Schneider, TL; Walsh, CT; O'Connor, SE. Utilization of alternate substrates by the first three modules of the epothilone synthetase assembly line. *J Am Chem Soc,* 2002 124, 11272-11273.

[227] Filipits, M; Pohl, G; Rudas, M; Dietze, O; Lax, S; Grill, R; Pirker, R; Zielinski, CC; Hausmaninger, H; Kubista, E; Samonigg, H; Jakesz, R. Clinical role of multidrug resistance protein 1 expression in chemotherapy resistance in early-stage breast cancer: the Austrian Breast and Colorectal Cancer Study Group. *J Clin Oncol,* 2005 23, 1161-1168.

[228] Bollag, DM; McQueney, PA; Zhu, J; Hensens, O; Koupal, L; Liesch, J; Goetz, M; Lazarides, E; Woods, CM. Epothilones, a new class of microtubule-stabilizing agents with a taxol-like mechanism of action. *Cancer Res,* 1995 55, 2325-2333.

[229] Mani, S; McDaid, H; Hamilton, A; Hochster, H; Cohen, MB; Khabelle, D; Griffin, T; Lebwohl, DE; Liebes, L; Muggia, F; Horwitz, SB. Phase I clinical and pharmacokinetic study of BMS-247550, a novel derivative of epothilone B, in solid tumors. *Clin Cancer Res,* 2004 10, 1289-1298.

[230] Kowalski, RJ; Giannakakou, P; Hamel E. Activities of the microtubule-stabilizing agents epothilones A and B with purified tubulin and in cells resistant to paclitaxel (Taxol(R)). *J Biol Chem,* 1997 Jan 24;272(4):2534-41.

[231] Altmann, KH; Wartmann, M; O'Reilly, T. Epothilones and related structures-a new class of microtubule inhibitors with potent in vivo antitumor activity. *Biochim Biophys Acta,* 2000 1470, M79-91.

[232] Bollag, DM; McQueney, PA; Zhu, J; Hensens, O; Koupal, L; Liesch, J; Goetz, M; Lazarides, E; Woods, CM. Epothilones, a new class of microtubule-stabilizing agents with a taxol-like mechanism of action. *Cancer Res,* 1995 55, 2325-2333.

[233] Bergstralh, DT; Taxman, DJ; Chou, TC; Danishefsky, SJ; Ting, JP. A comparison of signaling activities induced by Taxol and desoxyepothilone *B J Chemother,* 2004 16, 563-576.

[234] Goodin, S; Kane, MP; Rubin, EH. Epothilones: mechanism of action and biologic activity. *J Clin Oncol,* 2004 22, 2015-2025.

[235] Lee, FY; Borzilleri, R; Fairchild, CR; Kim, SH; Long, BH; Reventos-Suarez, C; Vite, GD; Rose, WC; Kramer, RA. BMS-247550: a novel epothilone analog with a mode of action similar to paclitaxel but possessing superior antitumor efficacy. *Clin Cancer Res,* 2001 7, 1429-1437.

[236] Altmann, KH. Recent developments in the chemical biology of epothilones. *Curr Pharm Des,* 2005 11, 1595-1613.

[237] Chou, TC; O'Connor, OA; Tong, WP; Guan, Y; Zhang, ZG; Stachel, SJ; Lee, C; Danishefsky, SJ. The synthesis, discovery, and development of

a highly promising class of microtubule stabilization agents: curative effects of desoxyepothilones B and F against human tumor xenografts in nude mice. *Proc Natl Acad Sci USA,* 2001 98, 8113-8118.

[238] Pazdur, R; Keegan, P. *FDA Approval for Ixabepilone.* National Cancer Institute, 2007 www.cancer.gov/cancertopics/druginfo/fda-ixabepilone.

[239] Rivera, E; Gomez, H. Chemotherapy resistance in metastatic breast cancer: the evolving role of ixabepilone. *Breast Cancer Res,* 2010 12 Suppl 2:S2.

[240] Jordan, MA; Kamath, K. How do microtubule-targeted drugs work? An overview. *Curr Cancer Drug Targets,* 2007 7, 730–742.

[241] Abrey, LE; Wen, P; Govindan, R; Reimers, H; Rigas, JR; Robins, HI; Allen-Freda, E; Gao, B; Ko J; Johri, A. Patupilone for the treatment of recurrent/progressive brain metastases in patients (pts) with non-small cell lung cancer (NSCLC): An open-label phase II study. J Clin Oncol, 2008 26 15 suppl:2033.

[242] Conlin, AK; Seidman, AD. Beyond cytotoxic chemotherapy for the first-line treatment of HER2-negative, hormone-insensitive metastatic breast cancer: current status and future opportunities. *Clin Breast Cancer,* 2008 8, 215-223.

[243] Hohmann, J; Molnár, J; Rédei, D; Evanics, F; Forgo, P; Kálmán, A; Argay, G; Szabó, P. Discovery and biological evaluation of a new family of potent modulators of multidrug resistance: reversal of multidrug resistance of mouse lymphoma cells by new natural jatrophane diterpenoids isolated from Euphorbia species. *J Med Chem,* 2002 45, 2425-2431.

[244] Ferreira, MJU; Gyémant, N; Madureira, AM; Tanaka, M; Koos, K; Didziapetris, R; Molnár, J. The effects of jatrophane derivatives on the reversion of MDR1- and MRP-mediated multidrug resistance in the MDA-MB-231 (HTB-26) cell line. *Anticancer Res,* 2005 25, 4173–4178.

[245] Aljancić, IS; Pesić, M; Milosavljević, SM; Todorović, NM; Jadranin, M; Milosavljević, G; Povrenović, D; Banković, J; Tanić, N; Marković, ID; Ruzdijić, S; Vajs, VE; Tesević, VV. Isolation and biological evaluation of jatrophane diterpenoids from Euphorbia dendroides. *J Nat Prod,* 2011 74, 1613-1620.

[246] Jadranin, M; Pešić, M; Aljančić, IS; Milosavljević, SM; Todorović, NM; Podolski-Renić, A; Banković, J; Tanić, N; Marković, I; Vajs, VE; Tešević, VV. Jatrophane diterpenoids from the latex of Euphorbia

dendroides and their anti-P-glycoprotein activity in human multi-drug resistant cancer cell lines. *Phytochemistry,* 2013 86, 208-217.
[247] Miglietta, A; Gabriel, L; Appendino, G; Bocca, C. Biological properties of jatrophane polyesters, new microtubule-interacting agents. *Cancer Chemother Pharmacol,* 2003 51, 67-74.
[248] Pešić, M; Banković, J; Aljančić, IS; Todorović, NM; Jadranin, M; Vajs, VE; Tešević, VV; Vučković, I; Momčilović, M; Marković, ID; Tanić, N; Ruždijić, S. New anti-cancer characteristics of jatrophane diterpenes from Euphorbia dendroides. *Food Chem Toxicol,* 2011 49, 3165-3173.
[249] Corea, G; Fattorusso, E; Lanzotti, V; Taglialatela-Scafati, O; Appendino, G; Ballero, M; Simon, P.N; Dumontet, C; Di Pietro, A. Jatrophane diterpenes as Pglycoprotein inhibitors. First insights of structure-activity relationships and discovery of a new, powerful lead. *J Med Chem,* 2003 46, 3395–3402.
[250] Pesic, M; Markovic, JZ; Jankovic, D; Kanazir, S; Markovic, ID; Rakic, L; Ruzdijic, S. Induced resistance in the human non small cell lung carcinoma (NCI-H460) cell line in vitro by anticancer drugs. *J Chemother,* 2006 18, 66-73.
[251] Belarbi el, H; Contreras Gómez, A; Chisti, Y; García Camacho, F; Molina Grima, E. Producing drugs from marine sponges. *Biotechnol Adv,* 2003 21, 585-598.
[252] Haefner, B. Drugs from the deep: marine natural products as drug candidates. *Drug Discov Today,* 2003 8, 536-544.
[253] Singh, R; Sharma, M; Joshi, P; Rawat, DS. Clinical status of anti-cancer agents derived from marine sources. *Anticancer Agents Med Chem,* 2008 8, 603-617.
[254] Sipkema, D; Franssen, MC; Osinga, R; Tramper, J; Wijffels, RH. Marine sponges as pharmacy. *Mar Biotechnol (NY),* 2005 7, 142-162.
[255] Fusetani, N; Kem, W. Marine toxins: an overview. *Prog Mol Subcell Biol,* 2009 46, 1-44.
[256] Pagea, MJ; Northcoteb, PT; Webbc, VL; Mackeyb, S; Handleya, SJ. Aquaculture trials for the production of biologically active metabolites in the New Zealand sponge Mycale hentscheli (Demospongiae: Poecilosclerida) *Aquaculture,* 2005 250, 256–269.
[257] Mooberry, SL; Tien, G; Hernandez, AH; Plubrukarn, A; Davidson, BS. Laulimalide and isolaulimalide, new paclitaxel-like microtubule-stabilizing agents. *Cancer Res,* 1999 59, 653-660.

[258] Clark, EA; Hills, PM; Davidson, BS; Wender, PA; Mooberry, SL. Laulimalide and synthetic laulimalide analogues are synergistic with paclitaxel and 2-methoxyestradiol. *Mol Pharm,* 2006 3, 457-467.
[259] Nogales, E; Medrano, FJ; Diakun, GP; Mant, GR; Towns-Andrews, E; Bordas, J. The effect of temperature on the structure of vinblastine-induced polymers of purified tubulin: detection of a reversible conformational change. *J Mol Biol,* 1995 254, 416-430.
[260] Downing, KH. Structural basis for the interaction of tubulin with proteins and drugs that affect microtubule dynamics. *Annu Rev Cell Dev Biol,* 2000 16, 89-111.
[261] Lu, H; Murtagh, J; Schwartz, EL. The microtubule binding drug laulimalide inhibits vascular endothelial growth factor-induced human endothelial cell migration and is synergistic when combined with docetaxel (taxotere). *Mol Pharmacol,* 2006 69, 1207-1215.
[262] Liu, J; Towle, MJ; Cheng, H; Saxton, P; Reardon, C; Wu, J; Murphy, EA; Kuznetsov, G; Johannes, CW; Tremblay, MR; Zhao, H; Pesant, M; Fang, FG; Vermeulen, MW; Gallagher, BM Jr; Littlefield, BA. In vitro and in vivo anticancer activities of synthetic (-)-laulimalide, a marine natural product microtubule stabilizing agent. *Anticancer Res* 2007 27, 1509-1518.
[263] Johnson, TA; Tenney, K; Cichewicz, RH; Morinaka, BI; White, KN; Amagata, T; Subramanian, B; Media, J; Mooberry, SL; Valeriote, FA; Crews, P. Sponge-derived fijianolide polyketide class: further evaluation of their structural and cytotoxicity properties. *J Med Chem,* 2007 50, 3795-3803.
[264] Mooberry, SL; Randall-Hlubek, DA; Leal, RM; Hegde, SG; Hubbard, RD; Zhang, L; Wender, PA. Microtubule-stabilizing agents based on designed laulimalide analogues. *Proc Natl Acad Sci USA,* 2004 101, 8803-8808.
[265] Long, BH; Carboni, JM; Wasserman, AJ; Cornell, LA; Casazza, AM; Jensen, PR; Lindel, T; Fenical, W; Fairchild, CR. Eleutherobin, a novel cytotoxic agent that induces tubulin polymerization, is similar to paclitaxel (Taxol). *Cancer Res,* 1998 58, 1111-1115.
[266] Longley, RE; Caddigan, D; Harmody, D; Gunasekera, M; Gunasekera, SP. Discodermolide--a new, marine-derived immunosuppressive compound. II. In vivo studies. *Transplantation,* 1991 52, 656-661.
[267] Gunasekera, SP; Longley, RE; Isbrucker, RA. Semisynthetic analogues of the microtubule-stabilizing agent discodermolide: preparation and biological activity. *J Nat Prod,* 2002 65, 1830-1837.

[268] Mani, S; Macapinlac, M Jr; Goel, S; Verdier-Pinard, D; Fojo, T; Rothenberg, M; Colevas, D. The clinical development of new mitotic inhibitors that stabilize the microtubule. *Anticancer Drugs,* 2004 15, 553-558.

[269] ter Haar, E; Kowalski, RJ; Hamel, E; Lin, CM; Longley, RE; Gunasekera, SP; Rosenkranz, HS; Day, BW. Discodermolide, a cytotoxic marine agent that stabilizes microtubules more potently than taxol. *Biochemistry,* 1996 35, 243-250.

[270] Martello, LA; LaMarche, MJ; He, L; Beauchamp, TJ; Smith, AB; Horwitz, SB. The relationship between Taxol and (+)-discodermolide: synthetic analogs and modeling studies. *Chem Biol,* 2001 8, 843-855.

[271] Martello, LA; McDaid, HM; Regl, DL; Yang, CP; Meng, D; Pettus, TR; Kaufman, MD; Arimoto, H; Danishefsky, SJ; Smith, AB 3rd; Horwitz, SB. Taxol and discodermolide represent a synergistic drug combination in human carcinoma cell lines. *Clin Cancer Res,* 2000 6, 1978-1987.

[272] Mita, A; Lockhart, AC; Chen, TL; Bochinski, K; Curtright, J; Cooper, W; Hammond, L; Rothenberg, M; Rowinsky E; Sharma, S. A phase I pharmacokinetic (PK) trial of XAA296A (Discodermolide) administered every 3 wks to adult patients with advanced solid malignancies. *Journal of Clinical Oncology,* 2004 22, 2004: 2025.

[273] Madiraju, C; Edler, MC; Hamel, E; Raccor, BS; Balachandran, R; Zhu, G; Giuliano, KA; Vogt, A; Shin, Y; Fournier, JH; Fukui, Y; Brückner, AM; Curran, DP; Day, BW. Tubulin assembly, taxoid site binding, and cellular effects of the microtubule-stabilizing agent dictyostatin. *Biochemistry,* 2005 44, 15053-15063.

[274] Canales, A; Matesanz, R; Gardner, NM; Andreu, JM; Paterson, I; Díaz, JF; Jiménez-Barbero, J. The bound conformation of microtubule-stabilizing agents: NMR insights into the bioactive 3D structure of discodermolide and dictyostatin. *Chemistry,* 2008 14, 7557-7569.

[275] Hood, KA; West, LM; Rouwé, B; Northcote, PT; Berridge, MV; Wakefield, SJ; Miller, JH. Peloruside A, a novel antimitotic agent with paclitaxel-like microtubule- stabilizing activity. *Cancer Res,* 2002 62, 3356-3360.

[276] Wilmes, A; Bargh, K; Kelly, C; Northcote, PT; Miller, JH. Peloruside A synergizes with other microtubule stabilizing agents in cultured cancer cell lines. *Mol Pharm,* 2007 4, 269-280.

[277] Gaitanos, TN; Buey, RM; Díaz, JF; Northcote, PT; Teesdale-Spittle, P; Andreu, JM; Miller, JH. Peloruside A does not bind to the taxoid site on

beta-tubulin and retains its activity in multidrug-resistant cell lines. *Cancer Res,* 2004 64, 5063-5067.
[278] Huzil, JT; Chik, JK; Slysz, GW; Freedman, H; Tuszynski, J; Taylor, RE; Sackett, DL; Schriemer, DC. A unique mode of microtubule stabilization induced by peloruside A. *J Mol Biol,* 2008 378, 1016-1030.
[279] Rowinsky, EK; Eisenhauer, EA; Chaudhry, V; Arbuck, SG; Donehower, RC. Clinical toxicities encountered with paclitaxel (Taxol). *Semin Oncol,* 1993 20, 1-15.
[280] Miller, JH; Singh, AJ; Northcote, PT. Microtubule-stabilizing drugs from marine sponges: focus on peloruside A and zampanolide. *Mar Drugs,* 2010 8, 1059-1079.
[281] Hirata, Y; Jemura, D. Halichondrins—antitumor polyether macrolides from a marine sponge. *Chem,* 1986 58, 701—710.
[282] Cigler, T; Vahdat, LT. Eribulin mesylate for the treatment of breast cancer. *Expert Opin Pharmacother,* 2010 11, 1587-1593.
[283] Kuznetsov, G; Towle, MJ; Cheng, H; Kawamura, T; TenDyke, K; Liu, D; Kishi, Y; Yu, MJ; Littlefield, BA. Induction of morphological and biochemical apoptosis following prolonged mitotic blockage by halichondrin B macrocyclic ketone analog E7389. *Cancer Res,* 2004 64, 5760-5766.

Reviewed by: Dr. Jelena Popović
Institute of Molecular Genetics and Genetic Engineering, Laboratory for Human Molecular Genetics, University of Belgrade, Vojvode Stepe 444a, Belgrade, Serbia
e-mail: jpopovic@imgge.bg.ac.rs

In: Paclitaxel
Editor: Diego Morales
ISBN: 978-1-62808-549-5

Chapter 2

Molecular Mechanisms that Mediate Resistance to Paclitaxel in Cancer

Rishil J. Kathawala, Charles R. Ashby Jr. and Zhe-Sheng Chen*
Department of Pharmaceutical Sciences, College of Pharmacy and Health Sciences, St. John's University, Queens, New York, NY, US

Abstract

Paclitaxel (Taxol) is an antimitotic drug that promotes microtubule polymerization and produces sensitization to radiation. Paclitaxel is a first-line drug for the treatment of non-small cell lung cancer and is used to treat metastatic ovarian, breast cancer, head and neck cancer, and advanced cases of Kaposi's sarcoma. The development of paclitaxel resistance, which reduces the efficacy of paclitaxel in a number of diseases, has been discussed in several review articles. Paclitaxel resistance can result from an increased efflux via ATP-binding cassette (ABC) multidrug transporters, changes in microtubule dynamics, genetic polymorphisms, alterations in extrinsic and intrinsic apoptotic pathway,

* Corresponding author, Zhe-Sheng Chen, Department of Pharmaceutical Sciences, College of Pharmacy and Health Sciences, St. John's University, Queens, NY 11439, USA. Tel, +1-718-990-1432; Email, chenz@stjohns.edu.

activation of transcription factors and altered target or binding sites. However, the pharmacokinetic and cellular resistance mechanisms of paclitaxel are not fully understood. Thus, elucidating paclitaxel resistance mechanisms would significantly facilitate the development of therapeutic interventions to surmount paclitaxel resistance. This chapter discusses the current knowledge about paclitaxel resistance at the molecular level.

Introduction

In 1971, paclitaxel isolation from the stem bark of pacific yew *Taxus brevifolia* was first reported by Mansukh Wani and colleagues at the Research Triangle Institute, North Carolina (Wani et al., 1971). However, it was not until 1993 that Bristol-Myers Squibb, in collaboration with National Cancer Institute (NCI), marketed synthetic paclitaxel under the trade name of Taxol (DeFuria and Horovitz, 1993). Paclitaxel, a cytotoxic compound that stabilizes microtubules, alone or in combination, is efficacious in treatment of number of cancers, including non-small cell lung cancer (Murphy et al., 1993), ovarian cancer (McGuire et al., 1996), breast cancer (Holmes et al., 1991), bladder cancer, head and neck cancer and a number of other malignancies such as Kaposi's sarcoma (Saville et al., 1995). Paclitaxel's novel mechanism of action provided hope that it would demonstrate anti-neoplastic efficacy in multidrug resistant tumor cells. Unfortunately, as with other chemotherapeutic drugs, resistance is commonly seen with paclitaxel treatment (Rao et al., 1994; Rao et al., 1995; Pinedo et al., 1996). Optimizing the uses of paclitaxel has been a challenge among clinicians and scientists. In-depth knowledge about paclitaxel resistance will not only facilitate effective use of this drug but also guide the development of new drugs. This chapter gives a brief overview on the current understanding of the mechanisms that produce paclitaxel resistance.

1. Cellular Mechanisms of Action of Paclitaxel

Paclitaxel exerts its cytotoxic effects by different mechanisms. It binds to β-tubulin in microtubules (Horwitz, 1994; Rao et al., 1994; Rao et al., 1995), causing the formation of stable polymerized microtubules (Schiff et al., 1979; Jordan et al., 1993; Jordan et al., 1996; Fan, 1999; Orr et al., 2003). The paclitaxel-induced mitotic arrest triggers the mitotic spindle checkpoint and results in apoptosis (Zaffaroni et al., 1998). Paclitaxel-mediated apoptosis depends on the concentration of paclitaxel used, the duration of exposure and the cell line being studied. Paclitaxel, at subnanomolar concentrations, augments the polymerization of microtubules by reducing the critical concentration of tubulin dimers, guanosine triphosphate(GTP) and microtubule associated - proteins (MAP) needed to form microtubules (Jordan et al., 1993; Fan, 1999). At low concentrations, it also affects interphase microtubules and genes mediating apoptosis (Jordan et al., 1996; Fan, 1999). At higher concentrations, paclitaxel shifts the equilibrium from microtubule dimers to polymers, and prevents disassociation under conditions of stress such as cold and exposure to calcium chloride (Schiff et al., 1979). Other mechanisms by which paclitaxel mediates apoptosis includes activation of the cell division control-2 (cdc-2) kinase, stabilization of cyclin B-1, activation of spindle checkpoint during metaphase to anaphase transition (Rudner and Murray, 1996); phosphorylation of Bcl-2 (Rudner and Murray, 1996; Ferlini et al., 2003); induction of IL -1β and tumor necrosis factor-α (TNF-α) (Beyaert and Fiers, 1994; Hieber and Heim, 1994). Alternate anti-tumor mechanisms of paclitaxel includes decrease in the invasive potential of tumor cells, by decreasing production of metalloproteinase in selected tumor cell lines (Stearns and Wang, 1992), inhibition of angiogenesis by inhibition of endothelial cell proliferation and inhibition of motility (Belotti et al., 1996).

2. Mechanisms that Produce Resistance to Paclitaxel

Paclitaxel resistance due to acquired (host factors) or intrinsic (genetic or epigenetic) mechanisms in tumor cells eventually occurs, giving rise to progressive chemotherapeutic failure and the developing of malignant tumors (Orr et al., 2003; Mielke, 2007; Gillet and Gottesman, 2010; Tiwari et al., 2011; Sodani et al., 2012; Siegel et al., 2013).

The efficacy of paclitaxel can be attenuated by the overexpression of multidrug efflux transporters (Gottesman, 2002; Chen et al., 2003; Hopper-Borge et al., 2011; Sodani et al., 2012), altered metabolism, decreased sensitivity to apoptosis (Blagosklonny and Fojo, 1999; Orr et al., 2003), alterations in microtubule dynamics, diminished interaction of paclitaxel to its cellular target (Dumontet and Sikic, 1999; Drukman and Kavallaris, 2002) and genetic polymorphisms (Yin et al., 2010). These aforementioned mechanisms of resistance typically produce chemotherapeutic failure.

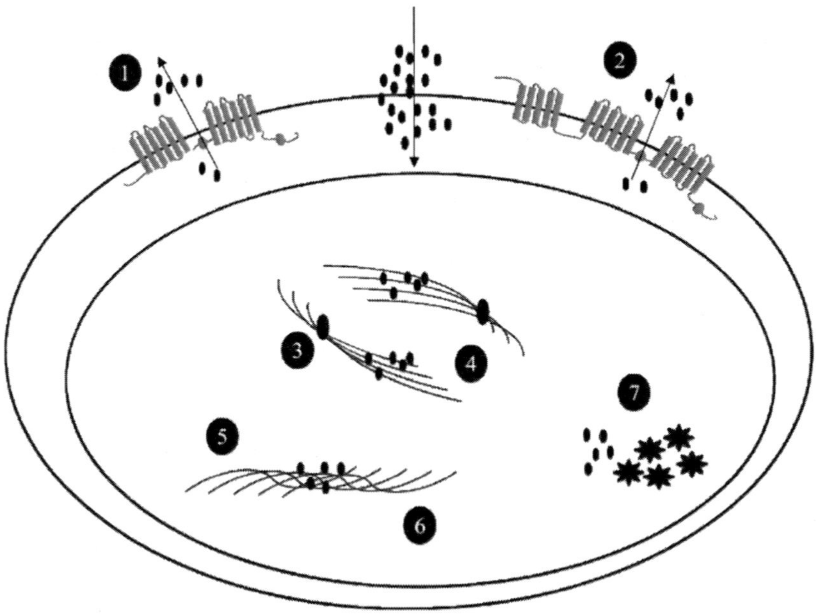

Kathawala et al.
Paclitaxel resistance can result from 1) increased efflux by ABCB1, 2) increased efflux by ABCC10, 3) alterations in α- and β-tubulin dynamics, paclitaxel binding site, mutations at β-tubulin, 4) post-translational modifications of tubulin and altered expression/modifications of MAPs, 5) alterations in Bcl-2, Bcl-xL, Bad, Bax, p53 and other genes, 6) modification of transcription factors like NAC-1, FoxM1, 7) apoptotic to autophagic switch, complete absence of caspase-mediated apoptotic cell death (using the pan-caspase-inhibitor Z-VAD).

2.1. Augmented Efflux by ABC Transporters

The ABC transporters are the most abundant transmembrane protein family encoded in the human genome (Dean et al., 2001; Ambudkar et al., 2006). The ATP-Binding Cassette (ABC) transporter family is divided into seven subfamilies, ABCA through ABCG (Gillet et al., 2007; Liu, 2009). Currently, 48 members of the ABC transporter family have been isolated and identified (Gillet et al., 2007; Wu et al., 2008), of which ABCB1 (MDR1, also called P-glycoprotein or P-gp) and ABCC10 (also called multidrug resistance protein 7 or MRP7) have been well characterized in terms of their capacities to confer MDR to paclitaxel (Alvarez et al., 1995; Schinkel et al., 1997; Chen et al., 2003; Hopper-Borge et al., 2004; Kruh et al., 2007; Oguri et al., 2008; Hopper-Borge et al., 2011; Sodani et al., 2012). The bile salt export protein ABCB11, first reported as a sister of ABCB1, confers a low level resistance to paclitaxel in transfected cells (Childs et al., 1998). The transporter-ABCB4 (MDR3), which is closely related to ABCB1, transports paclitaxel, although inefficiently unless it contains certain mutations (Smith et al., 2000; Gottesman et al., 2002; Schinkel and Jonker, 2003). However, the role of this transporter in paclitaxel resistance is unknown. There is no conclusive evidence to suggest that other potential drug efflux pumps such as ABCC1 (MRP1), lung resistance protein (LRP) and ABCG2 (also called breast cancer resistance protein or BCRP) play a role in paclitaxel resistance (Cole et al., 1994; Zaman et al., 1994; Huang et al., 1997a; Lee et al., 2000; Litman et al., 2000).

2.1.1. ABCB1/MDR1/P-gp Transporter

The human ABCB1 transporter, an *ABCB1* encoded gene product localized to chromosome 7p21, was the first identified ABC transporter (Juliano and Ling, 1976; Ueda et al., 1986). The ABCB1 transporter has a molecular weight of 170-kDa (Borst and Schinkel, 1997) and consists of two transmembrane binding domains (TMD1 and TMD2) and two nucleotide binding domains (NBD1 and NBD2) (Sauna et al., 2001; Gottesman et al., 2002). The ABCB1 is an apical membrane transporter that is located in the kidney, placenta, liver, adrenal glands, intestine and blood-brain barrier cells, where it functions to protect cells against toxic xenobiotics and cellular toxicants (Sauna et al., 2001; Gottesman et al., 2002; Sarkadi et al., 2006). The overexpression of ABCB1 confers significant resistance to a wide variety of neutral and cationic hydrophobic chemotherapeutic substrates, including taxanes (paclitaxel, docetaxel), epipodophyllotoxins (etoposide, teniposide), vinca alkaloids (vinblastine, vincristine), anthracyclines (doxorubicin, daunorubicin) (Gottesman et al., 2002; Schinkel and Jonker, 2003; Tiwari et al., 2011; Sodani et al., 2012), antibiotics such as dactinomycin and actinomycin D (Liu et al., 2001; Sauna et al., 2001; Findling-Kagan et al., 2005), tyrosine kinase inhibitors (TKIs) such as BCR-Abl TKIs like imatinib (Peng et al., 2011), nilotinib (Mahon et al., 2008), epidermal growth factor receptor (EGFR) TKI like erlotinib (Marchetti et al., 2008). Paclitaxel accumulates in the brain and gastrointestinal tract of ABCB1 knockout mice, indicating that ABCB1 prevents paclitaxel from crossing the blood-brain barrier and prevents biliary elimination (Schinkel et al., 1997). The evaluation of ABCB1 expression in the National Cancer Institute (NCI) 60 cell lines anticancer drug screening panel, by quantitative polymerase chain reaction (PCR), demonstrated a significant negative correlation coefficient (-0.896) for ABCB1 expression with the sensitivity profile of paclitaxel (Alvarez et al., 1995). There is considerable evidence indicating increased levels of either *ABCB1* mRNA (Schondorf et al., 1999; Schondorf et al., 2002) or ABCB1 protein (Parekh et al., 1997) in paclitaxel resistant cell lines. Therefore, it becomes important to identify the mechanisms that produce ABCB1 overexpression as these could also mediate paclitaxel resistance. Paclitaxel resistance could result from the overexpression of ABCB1, which may be due to amplification of the *ABCB1* gene (Schondorf et al., 2002), increased transcription of the *ABCB1* gene mediated by novel transcription factors such as RGP8.5 (Xu et al., 2001), mechanisms involving changes in ABCB1 translational efficiency (Schondorf et al., 2002), mutations in the *ABCB1* gene (Choi et al., 1988; Gros et al., 1991; Devine et al., 1992) and chromosomal rearrangements of the *ABCB1* gene resulting in the formation of hybrid genes (Mickley et al., 1997).

Clinically, the overexpression of ABCB1 has been shown in ovarian carcinoma (Bell et al., 1985; van der Zee et al., 1995), small-cell lung carcinoma (Holzmayer et al., 1992), breast carcinoma (Verrelle et al., 1991), acute leukemia (Sato et al., 1990; Pirker et al., 1991; De Moerloose, 2005), lymphoma (Wilson et al., 1995; Press et al., 1998), myeloma (Musto et al., 1991), retinoblastoma (Chan et al., 1991b), neuroblastoma (Chan et al., 1991a; De Moerloose, 2005), childhood sarcoma (Chan et al., 1990) after chemotherapy. It had been reported that there is a significant negative correlation between the level of ABCB1 expression in tumors and the response to paclitaxel (Bell et al., 1985; van der Zee et al., 1995). However, it had also been reported that there is no significant correlation between the expression of ABCB1 in ovarian carcinoma and response to chemotherapy (Arts et al., 1999; Katsaros et al., 1999). These conflicting data may be due to the use of different methods for detecting and quantitating ABCB1 levels (Robert, 1999).

There is an inconsistent pharmacogenetic correlation between paclitaxel and the ABCB1 transporter. The systemic elimination of paclitaxel occurs by hepatic metabolism involving the CYP450 enzymes CYP3A4 and CYP2C8 (Walle et al., 1995) and biliary elimination of paclitaxel occurs by ABCB1 (Sparreboom et al., 1997). Several single nucleotide polymorphisms (SNP) in the *ABCB1* gene, including G1199T/A and the linked G2677T/A

(Ala893Ser/Thr), are associated with progression-free survival after paclitaxel treatment (Green et al., 2006; Green et al., 2008) and C3435T (Ile1145Ile, wobble) as well as with paclitaxel-mediated peripheral neuropathy and neutropenia (Sissung et al., 2006). However, in other studies, no significant correlation was found between the pharmacokinetics of paclitaxel and the *ABCB1* genotype (Henningsson et al., 2005; Nakajima et al., 2005; Sissung et al., 2006; Grimm et al., 2010). In conclusion, there is no significant evidence indicating that genetic variations in *ABCB1* gene are associated with resistance to paclitaxel and altered pharmacokinetics. It has been hypothesized that *ABCC2* and *ABCG2* genetic variants may alter the therapeutic response to paclitaxel. However, to date, this hypothesis has not been verified (Marsh et al., 2007).

2.1.2. ABCC10/MRP7 Transporter

The human ABCC10 transporter, an *ABCC10* encoded gene product, is localized to chromosome 6p12 (Hopper et al., 2001; Kruh et al., 2007). The ABCC10 transporter is a 171-kDa protein containing three membrane spanning domains (MSD1, MSD2 and MSD3) and two NBDs. It belong to the class of long ABCCs, such as ABCC1, ABCC2, ABCC3, ABCC6 (Hopper et al., 2001; Chen et al., 2003; Sodani et al., 2012) and is localized to the basolateral cell surface (Malofeeva et al., 2012). The technique of reverse transcription-PCR analysis has shown a low level of *ABCC10* transcript expression in the skin, testes, spleen, stomach, colon, kidney, heart and brain (Hopper et al., 2001). It has been reported that *ABCC10* transcript expression was highest in the pancreas, followed by the liver, placenta, lungs, kidneys, brain, ovaries, lymph nodes, spleen, heart, leukocytes, colon (Takayanagi et al., 2004). ABCC10 is an amphipathic anion transporter whose physiological functions remain to be determined (Malofeeva et al., 2012). The substrates for ABCC10 are restricted to glucuronides such as 17-β-D-glucuronide ($E_2 17\beta G$) and the glutathione conjugate leukotriene C4 (LTC4) (Chen et al., 2003). The transfection of HEK293 cells with the *ABCC10* gene confers resistance to various antineoplastic drugs, including paclitaxel, docetaxel, vincristine, vinblastine, vinorelbine, cytarabine, gemcitabine, 2',3'-dideoxycytidine, 9-(2-phosphonyl methoxyethyl)adenine (PMEA), epothilone B (Hopper et al., 2001; Chen et al., 2003; Hopper-Borge et al., 2004; Kruh et al., 2007; Oguri et al., 2008; Bessho et al., 2009; Hopper-Borge et al., 2009; Hopper-Borge et al., 2011; Hu et al., 2011; Rudin et al., 2011; Malofeeva et al., 2012; Sun et al., 2013). In one study, it was shown that *Abcc10* knockout mice exhibited higher lethality due to neutropenia and marked bone marrow toxicity after treatment with paclitaxel (Hopper-Borge et al., 2011). This suggests that ABCC10 is required to protect cells against paclitaxel toxicity. With the exception of ABCC2 (Huisman et al., 2005) and ABCC10 (Sodani et al., 2012) no other ABCC family members have been reported to confer resistance to paclitaxel. ABCC10 expression has been reported in salivary gland adenocarcinoma and non-small cell lung carcinoma (Naramoto et al., 2007; Oguri et al., 2008; Bessho et al., 2009). In addition, *ABCC10* transcript has been detected in the HepG2 liver cancer cell line and two prostate cancer cell lines (CWR22Rv1 and TSUPR1) (Dabrowska and Sirotnak, 2004), as well as in breast, lung, colon, prostate, ovarian, and pancreatic tumor specimens (Takayanagi et al., 2004). In a recent study, chemosensitive and chemoresistant ovarian tumors in mice showed that ABCC10 and ABCB1 gene expression is increased by intermittent docetaxel treatment. This implies that the chemotherapy dosing schedule affects the development of resistance (De Souza et al., 2011).

It has been reported that ABCC10 expression is high in non-small cell lung cancer (Wang et al., 2009) and acute myeloid leukemia (Hu et al., 2011) samples after chemotherapy as a result of acquired MDR. Chen and colleagues have shown that various ABCC10 inhibitors such as cepharanthine (Zhou et al., 2009), imatinib, nilotinib (Shen et al., 2009; Tiwari et al., 2013), erlotinib, lapatinib (Kuang et al., 2010), tariquidar (Sun et al., 2013) can re-sensitize *ABCC10* transfected HEK293 cells to paclitaxel. In the future, efforts should be directed towards finding potent and selective ABCC10 inhibitors as they may be useful in surmounting paclitaxel resistance.

Currently, there are no published data regarding the pharmacogenetic correlation between paclitaxel and ABCC10. However, it has been reported that two ABCC10 SNPs (rs9349256 and rs2125739) and their haplotypes are significantly correlated with kidney tubular dysfunction, resulting in accelerated efflux of tenofovir by ABCC10 (Pushpakom et al., 2011). In addition, it has been suggested that the ABCC10 SNP rs2125739 is correlated with nevirapine-induced hepatotoxicity (Liptrott et al., 2012), although this has been disputed (Ciccacci et al., 2013).

2.2. Tubulin and Paclitaxel Resistance

Tubulin is a heterodimer that is composed of α and β tubulin subunits (Downing and Nogales, 1998; Nogales et al., 1998). Microtubules are dynamic and exhibit a nonequilibrium behavior termed dynamic instability where they undergo rapid stochastic transitions between growth and shrinkage, due to the association and dissociation of tubulin, respectively (Desai and Mitchison, 1997). The transition from growing to shrinking is termed "a catastrophe", whereas the reverse behavior is known as "a rescue" (Orr et al., 2003). Each β–tubulin molecule at its exchangeable or E-site binds to GTP, which leads to polymerization. The hydrolysis of GTP to GDP allows the microtubules to depolymerize (Orr et al., 2003). In the presence of paclitaxel, the hydrolysis of GTP to GDP occurs as usual but subsequent depolymerization is prevented (Rao et al., 1992), promoting microtubule stability (Snyder et al., 2001). α and β tubulin have three functional domains and a central domain (residues 207–384) composed of mixed β sheets (S7–S10) and three α helices (H8–H10), which connects α and β tubulin monomers longitudinally and laterally. Paclitaxel binds to a hydrophobic pocket within this central domain (Rao et al., 1992; Snyder et al., 2001; Orr et al., 2003). Below we will discuss how alterations in microtubule expression and dynamics produce resistance to paclitaxel.

2.2.1. Alterations in Microtubule Dynamics

The paclitaxel binding site is present only on polymerized tubulin. Consequently, increased microtubule dynamic instability, will increase the survival rate for a tumor challenged with paclitaxel (Cabral et al., 1986; Cabral and Barlow, 1989; Minotti et al., 1991; Wilson and Jordan, 1995; Goncalves et al., 2001). In addition, an increase in the incidence of paclitaxel resistance will increase the sensitivity of cells to substances that inhibit microtubule polymerization such as vinblastine and colchicine (Orr et al., 2003). In one study, the dynamics of individual rhodamine-labeled microtubules in paclitaxel-sensitive and -resistant A549 (human lung carcinoma cell line) were, quantified using digital time-lapse microscopy (Goncalves et al., 2001). The continuous exposure of the parental cell lines A549-T12 and -T24 to increasing concentrations of paclitaxel produced cell that were, 9- and 17-fold more resistant, respectively, to paclitaxel. Significantly, both resistant cell lines were dependent on low concentrations of paclitaxel (2 nM) for growth, and their division is inhibited in the G2/M phase of the cell cycle in the absence of paclitaxel. This suggests that both resistant cell lines exhibited increased dynamic instability compared with the parental cell line (Goncalves et al., 2001; Orr et al., 2003).

2.2.2. Altered Expression of β-Tubulin Isotypes

There are 6 different β-tubulin isotypes identified in humans, which includes class I β-tubulin encoded by the HM40 gene, class II by the Hβ9 gene, class III by the Hβ4 gene, class IVa by the Hβ5 gene, class IVb by the Hβ2 gene and class VI by the Hβ1 gene (Lewis et al., 1985; Lewis et al., 1987; Panda et al., 1994). However, only class I and IVb are constitutively expressed while other isotypes have tissue specific expression (Lewis et al., 1987). Different β–tubulin isotypes interact with different MAP and their differential expression also affects microtubule dynamics (Banerjee et al., 1992). Hence, inherent functional differences and differential isotype expression of β- tubulin could produce paclitaxel resistance (Orr et al., 2003; Yusuf et al., 2003). There is evidence that supports and refutes this hypothesis.

The overexpression of the Hβ2 tubulin gene produces cells resistant to paclitaxel (Haber et al., 1995). In one study, Zhou and colleagues demonstrated a 1.2-fold increase in Hβ2 tubulin and a 0.5-fold increase in Hβ3 tubulin levels in the QGY-TR50 cell line, which was 250-fold resistant to paclitaxel compared to the parent human hepatocellular carcinoma cell line (Zhou et al., 2001). Similarly, Ranganathan reported an increase in the Hβ3 isotype in the DU-145 paclitaxel resistant cell line (Ranganathan et al., 1998). The primary criticism of these studies is that they do not define whether tubulin overexpression is the primary cause of the paclitaxel resistance or is simply a secondary genetic change in cell lines chronically exposed to high concentrations of paclitaxel (Yusuf et al., 2003). There is also data negating the role of differential β-tubulin isotype expression in paclitaxel resistance. Lu and colleagues, using a monoclonal antibody against the Hβ3 isotype, showed that the assembly of Hβ3 depleted tubulin was increased relative to unfractionated tubulin, at different paclitaxel concentrations (Lu and Luduena, 1993). Similarly, Ranganathan and colleagues were unable to demonstrate paclitaxel resistance in human prostate carcinoma cell lines stably transfected with the cDNA of the Hβ3 isotype of tubulin (Ranganathan et al., 2001). Several studies have demonstrated increases in the levels of Hβ4 and H5β tubulin isotypes in paclitaxel resistant cell lines. Kavallaris reported that the A549-T12 and A549-T24 paclitaxel resistant cell lines (grown in 12 μM and 24 μM paclitaxel, respectively) significantly overexpressed Hβ4 gene (Kavallaris et al., 1997). In another study, Jaffrezou showed that in the KPTA5 cell line derived from the K562 leukemia cell line by selection using stepwise increases of paclitaxel and cyclosporin D, there was a 2- fold increase in the level of H5β tubulin mRNA and its corresponding class IVa β-tubulin protein

(Jaffrezou et al., 1995). In contrast to the findings described above, Dumontet and coworkers reported no significant association between an increase in Hβ4 and H5β tubulin isotypes and resistance to paclitaxel in a resistant human sarcoma cell line (Dumontet et al., 1996). Nicoletti et al. (2001) showed a significant correlation between the increased expression of the Hβ4 isotype of β-tubulin and paclitaxel resistance in 17 cell lines with little or no ABCB1 expression. However, no significant correlation could be demonstrated between isotype expression and paclitaxel resistance in 12 human ovarian carcinoma xenografts, established from patients samples taken before or after the start of paclitaxel-based therapy (Nicoletti et al., 2001). These inconsistencies can be attributed to most of the *in vitro* systems that evaluated paclitaxel resistance in cells that have a hundred-fold resistance to paclitaxel and have been selected by growing cells continuously in micromolar concentrations of paclitaxel. Thus, there is a possibility that alterations in tubulin are a result of *in vitro* selection methods and not reflective of clinical drug resistance (Yusuf et al., 2003). It can be argued that clinical exposure to paclitaxel is of much shorter duration and typically at lower paclitaxel doses. In one clinical study, Kavallaris and coworkers used highly specific oligonucleotides and PCR to compare β-tubulin isotypes between untreated primary and paclitaxel resistant ovarian epithelial tumors. Their results showed a significant increase in the levels of HM40, Hβ4 and Hβ5 β-tubulin in paclitaxel resistant tumor samples (Kavallaris et al., 1997).

The α-tubulin isotype composition also has the potential to affect the drug sensitivity of cells *in vitro*. The lung carcinoma cell line, NCI-H460/T800, which is 1000-fold resistant to paclitaxel compared to the parental cell line, overexpresses α-tubulin protein, but not the mRNA for this protein (Han et al., 2000). The downregulation of k-α1-tubulin in this resistant cell line, using an anti-sense DNA construct, produced a 45–51% increase in the sensitivity to paclitaxel in three independent clones. However, there was no evidence indicating that α-tubulin conferred resistance to paclitaxel (Orr et al., 2003).

2.2.3. Mutations of β-Tubulin

Paclitaxel- dependent β-tubulin mutant in Chinese hamster ovary (CHO) cell line is reported to have mutations in the region of the gene coding for amino acids 250-300 (Cabral et al., 1983). Drug sensitive cells that have been transformed into drug resistant cells by DNA from colcemid resistant cells contain a mutation in the β-tubulin gene (Whitfield et al., 1986). Subsequently, Giannakakou and coworkers, who established two paclitaxel resistant cell lines from the 1A9 clone of the A2780 ovarian carcinoma cell line, showed the presence of two point mutations in the HM40 isotype of β-tubulin, one at nucleotide 810 in one resistant clone and nucleotide 1092 in the other. Although these mutations were heterozygous at the genomic level, only the mutant allele of β-tubulin was expressed in both clones. These mutations were hypothesized to weaken the interaction between paclitaxel and β-tubulin (Giannakakou et al., 1997). There is controversy regarding how β-tubulin mutations confer paclitaxel resistance. Cabral did not find an alteration in drug binding in paclitaxel- resistant CHO cells with mutations in the β-tubulin gene. Mutations in β-tubulin may actually confer paclitaxel resistance by altering the dynamics of microtubule polymerization rather than altering paclitaxel binding to β-tubulin (Fujimura et al., 1992; James et al., 1993). This also explains why some of these mutations are not at the putative paclitaxel binding sites (Yusuf et al., 2003). Three epothilone (efficacious taxane congener) resistant cell lines selected from A549 and HeLa cells were crossresistant to all taxanes and did not express the ABCB1 gene (Orr et al., 2003). Sequence analysis of the class I β-tubulin from these resistant cell lines revealed single point mutations at β292 (Gln to Glu), β173 (Pro to Ala) and β422 (Tyr to Tyr/Cys), respectively. These mutations are near the M-loop, the nucleotide-binding site and the C-terminus regions that are involved in stabilizing the lateral contacts between adjacent protofilaments, the hydrolysis of GTP and the binding of MAPs, respectively. It is likely that these mutations decrease the endogenous stability of the microtubule to compensate for the activities of microtubule- stabilizing drugs. Consistent with this hypothesis, these resistant cell lines became more sensitive to microtubule-destabilizing drugs such as vinblastine and colchicine (Orr et al., 2003). The sequencing of the class I β-tubulin gene in theA549-T12 cells did not reveal any mutations. However, a heterozygous point mutation in K-α1 tubulin was found at residue 379 (Ser to Ser/Arg) (Martello et al., 2003).

In one study, Monzo sequenced exons 1 and 4 of paraffin embedded NSCLC samples from 49 patients, all of whom had received paclitaxel in the course of their treatment. Nineteen β-tubulin mutations were identified in sixteen tumors prior to paclitaxel treatment. Except for two of these mutations, all were identified within exon 4. None of the patients with β- tubulin mutations had complete or partial responses to platinum and paclitaxel whereas 13 of 33 (39.4 %) without β-tubulin mutations had complete or partial responses. The median survival time was 3 months for the 16 patients with β-tubulin mutations and 10 months for the 33 patients without β-tubulin mutations ($p<0.0001$) (Monzo et al., 1999). In another study, Kelley selected

20 NSCLC primary tumor samples. Sequencing of the coding regions of the TUBB (HM40 gene) in primary tumor samples demonstrated polymorphisms only in 2 of 25 NSCLC tumor lines. The results of Kelley and Monzo raise the possibility that the mutations described by Monzo were base changes in pseudogenes (Kelley et al., 2001). Indeed, Lamendola using PCR-SSCP analysis of exon 4 of the HM40 β-tubulin gene in paired human samples from 29 patients with ovarian cancer, did not find any tubulin mutations in both pre-chemotherapy and post-chemotherapy samples (Lamendola et al., 2003). Overall, further evidence is required before any conclusions can be made about the association of β-tubulin mutations and paclitaxel resistance (Yusuf et al., 2003).

2.3. Post-Translational Modifications of Tubulin and Altered Expression/Modifications of Microtubule Associated Proteins

Several structural and motor microtubule associated proteins (MAPs), including MAP-2, kinesin and tau, interact under *in vitro* conditions with the microtubule cytoskeleton that is regulated by the level of polyglutamylation of the α- and β-tubulins (Boucher et al., 1994; Larcher et al., 1996; Bonnet et al., 2001). The levels of tubulin glutamylation and tubulin polyglutamylase activity are cell cycle-dependent (Bobinnec et al., 1998; Regnard et al., 1999). Although enzymatic activity is highest in the G2 phase, the level of glutamylated tubulin is highest in mitosis, suggesting a complex regulation involving both polyglutamylase and deglutamylase activities. The removal of phosphate from βIII tubulin by protein phosphatase 2A inhibited the MAP-2-stimulation of *in vitro* microtubule assembly (Khan and Luduena, 1996). However, there is little evidence that altered post translational modifications are a major determinant of cellular sensitivity towards paclitaxel (Orr et al., 2003). Mitotic centromere-associated kinesin (MCAK) is a member of the kinesin superfamily of MAP motor proteins (Ems-McClung and Walczak, 2010). MCAK is also known as kif2c and has been classified as a member of the Kinesin-13 subgroup of motor proteins (Lawrence et al., 2004). MCAK is located at the spindle poles (Maney et al., 1998; Ganguly et al., 2008) and its overexpression causes a 7-fold increase in microtubule detachment from centrosomes and antagonizes paclitaxel's action by increasing the frequency of microtubule detachment, thereby allowing normal spindle assembly and cell division (Ganguly et al., 2011). The combination of MCAK and vinblastine enhances the drug effect on microtubule detachment and thus cells become more sensitive to drug (Ganguly et al., 2011). Tau, an MAP, binds to the exterior as well as interior microtubule surfaces, in the same binding site as paclitaxel, and consequently it can compete with paclitaxel. Tau stabilizes microtubules in the same way as paclitaxel, but with greater reversibility (Kar et al., 2003). Tau overexpression enhances polymerization and diminishes cell flexibility (Dye et al., 1993). One study reported that patients with advanced breast cancer receiving paclitaxel and cisplatin had high tau protein levels and this was associated with a poor response to chemotherapy, while low tau protein expression was indicative of a significant response to chemotherapy (Shao et al., 2010). An interesting retrospective study revealed that high risk, early breast cancer patients treated with paclitaxel, did not confirm the predictive role of tau protein (Pentheroudakis et al., 2009). The authors studied the prognostic and predictive role of estrogen receptor (ER) mRNA, progesterone receptor mRNA and tau mRNA. The results indicated a higher sensitivity to paclitaxel in low tau RNA tumors was observed only in ER-negative patients. Thus, it is possible that paclitaxel treatment effects in low tau RNA patients receiving hormonal therapy (ER-positive) were attenuated by resistance to hormonal therapy. This suggests tau mRNA expression is induced by estrogen and correlates with ER expression. High tau expression may be a good prognostic factor; however, in one study, this correlation was observed only in ER-positive breast cancer tumors (Pentheroudakis et al., 2009; Pusztai et al., 2009; Shao et al., 2010).

Stathmin, a microtubule destabilizer, and MAP4, a microtubule stabilizer, are proteins that regulate the dynamics of cellular microtubules (Orr et al., 2003). Stathmin binds to tubulin dimers and stimulate microtubule catastrophes (Belmont and Mitchison, 1996; Cassimeris, 2002). This destabilizing activity is regulated by phosphorylation and is lost when stathmin is phosphorylated (Marklund et al., 1996; Horwitz et al., 1997). MAP4 is also regulated by phosphorylation (Chapin et al., 1995; Chang and Karin, 2001). MAP4 alters microtubule dynamics by increasing the rescue frequency, without affecting the catastrophe frequency. The phosphorylation of MAP4 results in a loss of this microtubule stabilizing activity. The overexpression /activation of stathmin and/or the downregulation/inactivation of MAP4 increase the dynamicity and decrease the stability of microtubules (Larsson et al., 1999). The protein levels of stathmin and MAP4 have been quantified in the paclitaxel-sensitive and -resistant A549 cell lines. Stathmin protein levels in the A549-T12 and -T24 resistant cell lines were increased 2- fold compared to the parental cell line. In the parental A549 cells, exposure to increasing concentrations of paclitaxel caused a shift from the nonphosphorylated, active form of stathmin to the phosphorylated, inactive protein. Significantly, no shift in phosphorylation was observed in the two paclitaxel-resistant cell lines. With regard to MAP4, the parental cell line expressed the nonphosphorylated and active form of the microtubule-stabilizing protein, whereas A549-T24 cells predominately expressed the phosphorylated and inactive form (Orr et al., 2003). Stathmin mRNA levels are known to be upregulated in breast carcinoma, acute leukemias, lymphomas and various carcinomas (Hanash et al., 1988; Nylander et al., 1995; Bieche et al., 1998; Curmi et al., 2000). A phase 1 clinical study of sequential doxorubicin/vinorelbine treatment indicated a partial correlation between the induction of p53 and decreased MAP4 expression in peripheral blood mononuclear cells and tumors (Bash-Babula et al., 2002). These changes in cancer cells reduce the microtubule-stabilizing potency of paclitaxel, thereby inducing resistance (Orr et al., 2003).

2.4. Other Genetic Changes Related to Paclitaxel Resistance

Several other gene mutations have been implicated in paclitaxel resistance, with considerable work focusing on the role of the tumor suppressor gene p53 and the apoptosis regulatory genes Bcl-2, Bcl-xL, Bax, Bad (Yusuf et al., 2003). The loss of functional p53 does not produce paclitaxel resistance (Yusuf et al., 2003). Indeed, there is evidence suggesting that the loss of functional p53 can increase sensitivity to paclitaxel. For example, Vasey and colleagues have demonstrated that when A2780 human ovarian carcinoma cells expressing wild type p53 were transfected with a dominant negative mutant p53, they did not acquire resistance to paclitaxel (Vasey et al., 1996). Cassinelli and coworkers have shown that a cisplatin resistant subline of the ovarian carcinoma cell line IGROV-1, with mutant p53, has increased sensitivity to paclitaxel. The authors concluded that this was associated with the loss of the p53-dependent post mitotic spindle checkpoint, leading to delayed apoptosis (Cassinelli et al., 2001). Other studies suggest an alternative role of functional p53 in paclitaxel sensitivity. For example, using NCI-60 cell lines, anti-cancer drug screen demonstrated no significant correlation between p53 status and sensitivity to paclitaxel (O'Connor et al., 1997). There are several explanations as to why the lack of functional p53 could lead to an increase in the response to paclitaxel. The lack of p53-induced G1 arrest can induce cells to reach the G2/M phase of the cell cycle and undergo subsequent mitotic arrest. Also, in the absence of functional p53, p21 cannot be upregulated and hence cannot induce the exit of cells from mitotic arrest (Wang et al., 2000). The lack of functional p53 can upregulate MAP4 levels. This has been shown to increase microtubule polymerization, paclitaxel binding and sensitivity to paclitaxel (Zhang et al., 1999). These *in vitro* data represents that paclitaxel is successful in treating tumors with p53 mutations (Baker et al., 1989; Reiss et al., 1992; Rose, 1993). Giannakakou examined the p53 status of three paclitaxel selected human ovarian carcinoma cell lines resistant to paclitaxel as a result of β-tubulin mutations. Two of these cell lines expressed functionally inactive p53 and the third cell line showed a decrease in the wild type-p53 mRNA (pseudo-null phenotype). The introduction of wild type p53, however, did not restore paclitaxel sensitivity. Although this study suggests that the loss of p53 may play a role in the development of resistance, it does not distinguish between direct p53 effects and secondary effects due to increased genomic instability in p53 mutant cells (Giannakakou et al., 2000). Hence, it is reasonable to conclude that a lack of p53 function and paclitaxel resistance is not correlative. However, mutated p53 is considered as a clinical marker for paclitaxel resistance (Rosell et al., 1995; Goff et al., 1998).

The overexpression of Bcl-2 and Bcl-xL has been shown to contribute to paclitaxel resistance (Ibrado et al., 1997; Sharma et al., 2000). For example, a 2.5- fold increase in Bcl-2 levels and a 3- fold increase in Bcl-xL levels were demonstrated in the paclitaxel-resistant human leukemia cell line HL-60/ TX1000. However, a very high expression of Bcl-2 (5-fold) and Bcl-xL (10-fold) was needed to confer paclitaxel resistance in Bcl-2 and Bcl-xL transfectants of this HL-60 cell line (Huang et al., 1997a). This suggests that there may be a certain threshold above which these anti apoptotic genes confer resistance to paclitaxel. Lower levels of overexpression may be due to the genetic instability produced by the processes that ultimately produce cancer (Yusuf et al., 2003). In another study involving the breast cancer cell line MCF-7, paclitaxel-induced lethal DNA fragmentation was observed only in the absence or in the presence of very low concentrations of estradiol in the culture medium. The cells that were cultured in a medium containing high estradiol level showed a significantly lower paclitaxel-induced loss of cell viability and did not increase when paclitaxel concentrations were increased above 0.05 μM, as compared to cells cultured in medium with very low estadiol concentrations. Estradiol, in this study, produced a 4-fold elevation in Bcl-2vbut had no significant effect on Bax levels (elevated ratio of Bcl 2/Bax). This supports the notion that elevated p26Bcl-2 may play a role in conferring paclitaxel resistance (Huang et al., 1997b). Srivastava examined the expression of Bcl-2 and Bax in the Bcl-2 positive MCF-7 and the Bcl-2 null MDA 231 breast carcinoma cell lines, both of which were transfected with Bcl-2. Paclitaxel was shown to induce Bcl-2 hyperphosphorylation and Bax expression, which produced apoptosis in these cell lines. The authors proposed that Bcl-2 hyperphosphorylation decreased the ability of the molecule to heterodimerize with Bax, resulting in the release of Bax in its free state. This coupled with the induction of Bax by paclitaxel, resulted in increased formation of Bax, Bax homodimers and subsequent apoptosis. Increased Bcl-2 levels were associated with protection from apoptosis in both MCF 7 and MDA 231 cells in the presence of low levels of paclitaxel (Srivastava et al., 1998). These findings are supported by studies demonstrating that forced Bax overexpression produced a significant increase in paclitaxel sensitivity after 48, 72 and 120 hours of exposure (Strobel et al., 1996). Similarly, co-transfection of the 36M and SW626 ovarian carcinoma cell lines with the *BAD* gene also induced an increase in sensitivity to paclitaxel. The overexpression of Bad also occurs in malignant ascites from ovarian carcinoma. Thus, Bad may be a physiological regulator of the response of some cells to paclitaxel (Yusuf et al., 2003). There is some evidence to suggest that Bcl-2 overexpression may not contribute to paclitaxel resistance (Ogretmen and Safa, 1996; Luo et al., 2000); however, one needs to consider that the interactions of members of the Bcl family of proteins are complex. Since chemotherapy-induced cell death occurs via an apoptotic pathway, it is not surprising that changes in the ratio of pro- and anti-apoptotic proteins should produce changes in the relative sensitivity paclitaxel (Yusuf et al., 2003). Other genetic markers of paclitaxel resistance have been identified, including mutated *K-RAS* (Rosell et al., 1995), increased levels of *HER-2/neu* (Colomer et al., 2000) and decreased *p27* expression (Goff et al., 2001).

2.5. Transcription Factors and Paclitaxel Resistance

NAC-1 is a protein coded by the *BTB/POZ* gene (Nakayama et al., 2006), that has been associated with carcinoma. NAC-1, is a transcription repressor and is involved in self-renewal and maintenance of pluripotency in embryonic stem cells (Wang et al., 2006). NAC-1 expression is higher in ovarian serous carcinoma specimens obtained after administration of chemotherapy (Davidson et al., 2007). The overexpression of NAC-1 is significantly correlated with shorter relapse-free survival in patients with advanced stage (stage III/IV) ovarian carcinoma treated with platinum and taxane chemotherapy (Ishibashi et al., 2008). The knockdown of the gene for NAC-1 in both paclitaxel-resistant human ovarian cancer cell lines KF28TX and KFr13TX restores the efficacy of paclitaxel (Ishibashi et al., 2008).

The proliferation-specific oncogenic transcription factor FoxM1 is overexpressed in a broad range of tumor types including those of mammary, neural, gastrointestinal, and reproductive origin (Nakamura et al., 2004; Pilarsky et al., 2004; Liu et al., 2006; Bektas et al., 2008). FoxM1 is regulated throughout the cell cycle by phosphorylation. Growth factors activate surface receptors to initiate signaling pathways such as phosphoinositide 3-kinase (PI3K)/Akt and Ras/Raf/mitogen-activated protein kinase which promote cell division (Le et al., 2005) which inturn activates kinases such as cyclin-cdk and polo-like kinase 1 (PLK1), resulting in the phosphorylation and activation of FoxM1 (Major et al., 2004; Fu et al., 2008; Chen et al., 2009). The overexpression or silencing of human epidermal growth factor receptor 2 (HER2)/ErbB2, a surface receptor, is significantly correlated with FoxM1 levels in mammary cell lines and in transgenic mouse models (Francis et al., 2009). In human mammary tumors, FoxM1 expression was reported to be 8.7-fold higher in tumor cells versus normal tissue controls. In addition, it was shown that FoxM1 expression is significantly correlated with that of HER2 and FoxM1 levels were predictive of poor patient outcome (Bektas et al., 2008). FoxM1, also alters microtubule dynamics through the regulation of stathmin and this renders cells resistant to paclitaxel treatment (Carr et al., 2010).

2.6. Cancer Stem Cells Markers and Related Signaling Pathways Associated with Paclitaxel Resistance

Cells that have enhanced tumor initiating potential are referred to as cancer stem cells (CSC) (Abdullah and Chow, 2013). The defining characteristics of CSCs include the ability to self-renew and differentiate into other tumor cell subtypes. The WNT/β-catenin signaling pathway, is required for normal stem and CSC self -renewal in a number of cell types (Reya et al., 2003; Zhao et al., 2007; Bisson and Prowse, 2009). In c-kit+ ovarian CSCs, chemoresistance to cisplatin and paclitaxel results from the overexpression of ABCG2. The ABCG2-induced chemoresistance to both cisplatin and paclitaxel can be reversed by β-catenin siRNA knockdown (Chau et al., 2012). The notch signaling pathway plays an important role in a number of processes during tumor progression and metastases including tumor initiation, angiogenesis, epithelial-mesenchymal transition driven metastatic growth and self-renewal of CSCs (Ranganathan et al., 2011). The notch signaling pathway also contributes to chemoresistance in CD133+ glioma CSCs in cooperation with the Sonic hedgehog (SHH) pathway (Ulasov et al., 2011; Lin and Matsui, 2012). The abnormal activation of the SHH pathway in CSC models has been reported to produce resistance to paclitaxel, cisplatin and temozolomide (Song et al., 2011). While the specific mechanisms are not clear, the inhibition of the SHH pathway could sensitize CSCs in gastric, pancreatic, ovarian and prostate cancers (Song et al., 2011; Yao et al., 2011; Singh et al., 2012; Steg et al., 2012). Regulators of inflammation, such as the NF-κB, could also contribute to chemoresistance. NF-κB has a diverse set of biological function that can contribute to both pro-tumorigenic and anti-tumorigenic responses (Perkins, 2012). The constitutive activation of NF-κB and other pro-inflammatory signals in CD44+ ovarian CSCs is significantly correlated with chemoresistance to paclitaxel and carboplatin (Alvero et al., 2009). The inhibition of NF-κB by eriocalyxin B produces cell death in chemoresistant CD44+ ovarian CSCs (Leizer et al., 2011). In breast CSCs, treatment with disulfiram and copper can inhibit consitutively active NF-κB while sensitizing these cells to paclitaxel (Yip et al., 2011).

Lin28, a marker of CSCs, contributes to tumor relapse after chemotherapy and its overexpression promotes cancer cell proliferation (Xu et al., 2009; Yang et al., 2010; Pan et al., 2011). Lin28 is highly expressed in some tumors, such as hepatocellular carcinoma (Guo et al., 2006) and breast cancer cells (Lv et al., 2012). The overexpression of Lin28 induces p21 and Rb expression and inhibits let-7 miRNA (Lv et al., 2012). The protein let-7 suppresses tumor formation in lung and breast cancer by repressing oncogenes such as Hmga2 (Lee and Dutta, 2007) and Ras (Johnson et al., 2005; Besancon et al., 2009), suggesting that disruption of let-7 processing by activation of Lin28 could promote the oncogenic phenotype. A chemosensitive state can be induced by let-7 miRNA overexpression *in vitro* in ovarian cancer cells, whereas decreasing let-7 levels causes chemoresistance (Boyerinas et al., 2012; Lv et al., 2012). Although let-7 is not a universal sensitizer of cancer cells to paclitaxel, it affects acquired resistance of cells to paclitaxel by targeting IMP-1, resulting in destabilization of the mRNA of ABCB1. Consistently, many relapsed ovarian cancer patients tested before and after chemotherapy were found to have lower levels of let-7 but an upregulation of IMP-1 and ABCB1. Furthermore, the increased expression of both proteins after chemotherapy was negatively correlated with the length of disease-free time before recurrence (Boyerinas et al., 2012).

Belotti, D., Vergani, V., Drudis, T., Borsotti, P., Pitelli, M. R., Viale, G., Giavazzi, R. & Taraboletti, G. (1996). The microtubule-affecting drug paclitaxel has antiangiogenic activity. *Clin Cancer Res*, *2*, 1843-1849.

Besancon, R., Valsesia-Wittmann, S., Puisieux, A., Caron de Fromentel, C. & Maguer-Satta, V. (2009). Cancer stem cells, the emerging challenge of drug targeting. *Curr Med Chem*, *16*, 394-416.

Bessho, Y., Oguri, T., Ozasa, H., Uemura, T., Sakamoto, H., Miyazaki, M., Maeno, K., Sato, S. & Ueda, R. (2009). ABCC10/MRP7 is associated with vinorelbine resistance in non-small cell lung cancer. *Oncol Rep*, *21*, 263-268.

Beyaert, R. & Fiers, W. (1994). Molecular mechanisms of tumor necrosis factor-induced cytotoxicity. What we do understand and what we do not. *FEBS Lett*, *340*, 9-16.

Bieche, I., Lachkar, S., Becette, V., Cifuentes-Diaz, C., Sobel, A., Lidereau, R. & Curmi, P. A. (1998). Overexpression of the stathmin gene in a subset of human breast cancer. *Br J Cancer*, *78*, 701-709.

Bisson, I. & Prowse, D. M. (2009). WNT signaling regulates self-renewal and differentiation of prostate cancer cells with stem cell characteristics. *Cell Res*, *19*, 683-697.

Blagosklonny, M. V. & Fojo, T. (1999). Molecular effects of paclitaxel, myths and reality (a critical review). *Int J Cancer*, *83*, 151-156.

Bobinnec, Y., Moudjou, M., Fouquet, J. P., Desbruyeres, E., Edde, B. & Bornens, M. (1998). Glutamylation of centriole and cytoplasmic tubulin in proliferating non-neuronal cells. *Cell Motil Cytoskeleton*, *39*, 223-232.

Bonnet, C., Boucher, D., Lazereg, S., Pedrotti, B., Islam, K., Denoulet, P. & Larcher, J. C. (2001). Differential binding regulation of microtubule-associated proteins MAP1A, MAP1B, and MAP2 by tubulin polyglutamylation. *J Biol Chem*, *276*, 12839-12848.

Borst, P. & Schinkel, A. H. (1997). Genetic dissection of the function of mammalian P-glycoproteins. *Trends Genet*, *13*, 217-222.

Boucher, D., Larcher, J. C., Gros, F. & Denoulet, P. (1994). Polyglutamylation of tubulin as a progressive regulator of in vitro interactions between the microtubule-associated protein Tau and tubulin. *Biochemistry*, *33*, 12471-12477.

Boyerinas, B., Park, S. M., Murmann, A. E., Gwin, K., Montag, A. G., Zillhardt, M., Hua, Y. J., Lengyel, E. & Peter, M. E. (2012). Let-7 modulates acquired resistance of ovarian cancer to Taxanes via IMP-1-mediated stabilization of multidrug resistance 1. *Int J Cancer*, *130*, 1787-1797.

Cabral, F. & Barlow, S. B. (1989). Mechanisms by which mammalian cells acquire resistance to drugs that affect microtubule assembly. *FASEB J*, *3*, 1593-1599.

Cabral, F., Wible, L., Brenner, S. & Brinkley, B. R. (1983). Taxol-requiring mutant of Chinese hamster ovary cells with impaired mitotic spindle assembly. *J Cell Biol*, *97*, 30-39.

Cabral, F. R., Brady, R. C. & Schibler, M. J. (1986). A mechanism of cellular resistance to drugs that interfere with microtubule assembly. *Ann N Y Acad Sci*, *466*, 745-756.

Carr, J. R., Park, H. J., Wang, Z., Kiefer, M. M. & Raychaudhuri, P. (2010). FoxM1 mediates resistance to herceptin and paclitaxel. *Cancer Res*, *70*, 5054-5063.

Cassimeris, L. (2002). The oncoprotein 18/stathmin family of microtubule destabilizers. *Curr Opin Cell Biol*, *14*, 18-24.

Cassinelli, G., Supino, R., Perego, P., Polizzi, D., Lanzi, C., Pratesi, G. & Zunino, F. (2001). A role for loss of p53 function in sensitivity of ovarian carcinoma cells to taxanes. *Int J Cancer*, *92*, 738-747.

Chan, H. S., Haddad, G., Thorner, P. S., DeBoer, G., Lin, Y. P., Ondrusek, N., Yeger, H. & Ling, V. (1991a). P-glycoprotein expression as a predictor of the outcome of therapy for neuroblastoma. *N Engl J Med*, *325*, 1608-1614.

Chan, H. S., Thorner, P. S., Haddad, G. & Gallie, B. L. (1991b). Multidrug-resistant phenotype in retinoblastoma correlates with P-glycoprotein expression. *Ophthalmology*, *98*, 1425-1431.

Chan, H. S., Thorner, P. S., Haddad, G. & Ling, V. (1990). Immunohistochemical detection of P-glycoprotein, prognostic correlation in soft tissue sarcoma of childhood. *J Clin Oncol*, *8*, 689-704.

Chang, L. & Karin, M. (2001). Mammalian MAP kinase signalling cascades. *Nature*, *410*, 37-40.

Chapin, S. J., Lue, C. M., Yu, M. T. & Bulinski, J. C. (1995). Differential expression of alternatively spliced forms of MAP4, a repertoire of structurally different microtubule-binding domains. *Biochemistry*, *34*, 2289-2301.

Chau, W. K., Ip, C. K., Mak, A. S., Lai, H. C. & Wong, A. S. (2012). c-Kit mediates chemoresistance and tumor-initiating capacity of ovarian cancer cells through activation of Wnt/beta-catenin-ATP-binding cassette G2 signaling. *Oncogene*.

Chen, Y. J., Dominguez-Brauer, C., Wang, Z., Asara, J. M., Costa, R. H., Tyner, A. L., Lau, L. F. & Raychaudhuri, P. (2009). A conserved phosphorylation site within the forkhead domain of FoxM1B is required for its activation by cyclin-CDK1. *J Biol Chem*, *284*, 30695-30707.

Chen, Z. S., Hopper-Borge, E., Belinsky, M. G., Shchaveleva, I., Kotova, E. & Kruh, G. D. (2003). Characterization of the transport properties of human multidrug resistance protein 7 (MRP7, ABCC10). *Mol Pharmacol*, *63*, 351-358.

Childs, S., Yeh, R. L., Hui, D. & Ling, V. (1998). Taxol resistance mediated by transfection of the liver-specific sister gene of P-glycoprotein. *Cancer Res*, *58*, 4160-4167.

Choi, K. H., Chen, C. J., Kriegler, M. & Roninson, I. B. (1988). An altered pattern of cross-resistance in multidrug-resistant human cells results from spontaneous mutations in the mdr1 (P-glycoprotein) gene. *Cell*, *53*, 519-529.

Ciccacci, C., Di Fusco, D., Marazzi, M. C., Liotta, G., Palombi, L., Novelli, G. & Borgiani, P. (2013). ABCC10 rs2125739 polymorphism and nevirapine-induced hepatotoxicity, lack of association in a population from Mozambique. *Pharmacogenet Genomics*, *23*, 38-39.

Cole, S. P., Sparks, K. E., Fraser, K., Loe, D. W., Grant, C. E., Wilson, G. M. & Deeley, R. G. (1994). Pharmacological characterization of multidrug resistant MRP-transfected human tumor cells. *Cancer Res*, *54*, 5902-5910.

Colomer, R., Montero, S., Lluch, A., Ojeda, B., Barnadas, A., Casado, A., Massuti, B., Cortes-Funes, H. & Lloveras, B. (2000). Circulating HER2 extracellular domain and resistance to chemotherapy in advanced breast cancer. *Clin Cancer Res*, *6*, 2356-2362.

Curmi, P. A., Nogues, C., Lachkar, S., Carelle, N., Gonthier, M. P., Sobel, A., Lidereau, R. & Bieche, I. (2000). Overexpression of stathmin in breast carcinomas points out to highly proliferative tumours. *Br J Cancer*, *82*, 142-150.

Dabrowska, M. & Sirotnak, F. M. (2004). Regulation of transcription of the human MRP7 gene. Characteristics of the basal promoter and identification of tumor-derived transcripts encoding additional 5' end heterogeneity. *Gene*, *341*, 129-139.

Davidson, B., Berner, A., Trope, C. G., Wang, T. L. & Shih Ie, M. (2007). Expression and clinical role of the bric-a-brac tramtrack broad complex/poxvirus and zinc protein NAC-1 in ovarian carcinoma effusions. *Hum Pathol*, *38*, 1030-1036.

De Moerloose, B. (2005). [The prognostic significance of P-glycoprotein in children with acute lymphoblastic leukemia and neuroblastoma]. *Verh K Acad Geneeskd Belg*, *67*, 45-54.

De Souza, R., Zahedi, P., Badame, R. M., Allen, C. & Piquette-Miller, M. (2011). Chemotherapy dosing schedule influences drug resistance development in ovarian cancer. *Mol Cancer Ther*, *10*, 1289-1299.

Dean, M., Rzhetsky, A. & Allikmets, R. (2001). The human ATP-binding cassette (ABC) transporter superfamily. *Genome Res*, *11*, 1156-1166.

DeFuria, M. D. & Horovitz, Z. (1993). Taxol commercial supply strategy. *J Natl Cancer Inst Monogr*, 195-198.

Desai, A. & Mitchison, T. J. (1997). Microtubule polymerization dynamics. *Annu Rev Cell Dev Biol*, *13*, 83-117.

Devine, S. E., Ling, V. & Melera, P. W. (1992). Amino acid substitutions in the sixth transmembrane domain of P-glycoprotein alter multidrug resistance. *Proc Natl Acad Sci U S A*, *89*, 4564-4568.

Downing, K. H. & Nogales, E. (1998). Tubulin and microtubule structure. *Curr Opin Cell Biol*, *10*, 16-22.

Drukman, S. & Kavallaris, M. (2002). Microtubule alterations and resistance to tubulin-binding agents (review). *Int J Oncol*, *21*, 621-628.

Dumontet, C., Duran, G. E., Steger, K. A., Beketic-Oreskovic, L. & Sikic, B. I. (1996). Resistance mechanisms in human sarcoma mutants derived by single-step exposure to paclitaxel (Taxol). *Cancer Res*, *56*, 1091-1097.

Dumontet, C. & Sikic, B. I. (1999). Mechanisms of action of and resistance to antitubulin agents, microtubule dynamics, drug transport, and cell death. *J Clin Oncol*, *17*, 1061-1070.

Dye, R. B., Fink, S. P. & Williams, R. C., Jr. (1993). Taxol-induced flexibility of microtubules and its reversal by MAP-2 and Tau. *J Biol Chem*, *268*, 6847-6850.

Ems-McClung, S. C. & Walczak, C. E. (2010). Kinesin-13s in mitosis, Key players in the spatial and temporal organization of spindle microtubules. *Semin Cell Dev Biol*, *21*, 276-282.

Fan, W. (1999). Possible mechanisms of paclitaxel-induced apoptosis. *Biochem Pharmacol*, *57*, 1215-1221.

Ferlini, C., Raspaglio, G., Mozzetti, S., Distefano, M., Filippetti, F., Martinelli, E., Ferrandina, G., Gallo, D., Ranelletti, F. O. & Scambia, G. (2003). Bcl-2 down-regulation is a novel mechanism of paclitaxel resistance. *Mol Pharmacol*, *64*, 51-58.

Findling-Kagan, S., Sivan, H., Ostrovsky, O., Nagler, A. & Galski, H. (2005). Establishment and characterization of new cellular lymphoma model expressing transgenic human MDR1. *Leuk Res*, *29*, 407-414.

Francis, R. E., Myatt, S. S., Krol, J., Hartman, J., Peck, B., McGovern, U. B., Wang, J., Guest, S. K., Filipovic, A., Gojis, O., Palmieri, C., Peston, D.,

Shousha, S., Yu, Q., Sicinski, P., Coombes, R. C. & Lam, E. W. (2009). FoxM1 is a downstream target and marker of HER2 overexpression in breast cancer. *Int J Oncol*, *35*, 57-68.

Fu, Z., Malureanu, L., Huang, J., Wang, W., Li, H., van Deursen, J. M., Tindall, D. J. & Chen, J. (2008). Plk1-dependent phosphorylation of FoxM1 regulates a transcriptional programme required for mitotic progression. *Nat Cell Biol*, *10*, 1076-1082.

Fujimura, M., Oeda, K., Inoue, H. & Kato, T. (1992). A single amino-acid substitution in the beta-tubulin gene of Neurospora confers both carbendazim resistance and diethofencarb sensitivity. *Curr Genet*, *21*, 399-404.

Ganguly, A., Bhattacharya, R. & Cabral, F. (2008). Cell cycle dependent degradation of MCAK, evidence against a role in anaphase chromosome movement. *Cell Cycle*, *7*, 3187-3193.

Ganguly, A., Yang, H. & Cabral, F. (2011). Overexpression of mitotic centromere-associated Kinesin stimulates microtubule detachment and confers resistance to paclitaxel. *Mol Cancer Ther*, *10*, 929-937.

Giannakakou, P., Poy, G., Zhan, Z., Knutsen, T., Blagosklonny, M. V. & Fojo, T. (2000). Paclitaxel selects for mutant or pseudo-null p53 in drug resistance associated with tubulin mutations in human cancer. *Oncogene*, *19*, 3078-3085.

Giannakakou, P., Sackett, D. L., Kang, Y. K., Zhan, Z., Buters, J. T., Fojo, T. & Poruchynsky, M. S. (1997). Paclitaxel-resistant human ovarian cancer cells have mutant beta-tubulins that exhibit impaired paclitaxel-driven polymerization. *J Biol Chem*, *272*, 17118-17125.

Gillet, J. P., Efferth, T. & Remacle, J. (2007). Chemotherapy-induced resistance by ATP-binding cassette transporter genes. *Biochim Biophys Acta*, *1775*, 237-262.

Gillet, J. P. & Gottesman, M. M. (2010). Mechanisms of multidrug resistance in cancer. *Methods Mol Biol*, *596*, 47-76.

Goff, B. A., Paley, P. J., Greer, B. E. & Gown, A. M. (2001). Evaluation of chemoresistance markers in women with epithelial ovarian carcinoma. *Gynecol Oncol*, *81*, 18-24.

Goff, B. A., Ries, J. A., Els, L. P., Coltrera, M. D. & Gown, A. M. (1998). Immunophenotype of ovarian cancer as predictor of clinical outcome, evaluation at primary surgery and second-look procedure. *Gynecol Oncol*, *70*, 378-385.

Goncalves, A., Braguer, D., Kamath, K., Martello, L., Briand, C., Horwitz, S., Wilson, L. & Jordan, M. A. (2001). Resistance to Taxol in lung cancer cells associated with increased microtubule dynamics. *Proc Natl Acad Sci U S A*, *98*, 11737-11742.

Gottesman, M. M. (2002). Mechanisms of cancer drug resistance. *Annu Rev Med*, *53*, 615-627.

Gottesman, M. M., Fojo, T. & Bates, S. E. (2002). Multidrug resistance in cancer, role of ATP-dependent transporters. *Nat Rev Cancer*, *2*, 48-58.

Green, H., Soderkvist, P., Rosenberg, P., Horvath, G. & Peterson, C. (2006). mdr-1 single nucleotide polymorphisms in ovarian cancer tissue, G2677T/A correlates with response to paclitaxel chemotherapy. *Clin Cancer Res*, *12*, 854-859.

Green, H., Soderkvist, P., Rosenberg, P., Horvath, G. & Peterson, C. (2008). ABCB1 G1199A polymorphism and ovarian cancer response to paclitaxel. *J Pharm Sci*, *97*, 2045-2048.

Grimm, C., Polterauer, S., Zeillinger, R., Tong, D., Heinze, G., Wolf, A., Natter, C., Reinthaller, A. & Hefler, L. A. (2010). Two multidrug-resistance (ABCB1) gene polymorphisms as prognostic parameters in women with ovarian cancer. *Anticancer Res*, *30*, 3487-3491.

Gros, P., Dhir, R., Croop, J. & Talbot, F. (1991). A single amino acid substitution strongly modulates the activity and substrate specificity of the mouse mdr1 and mdr3 drug efflux pumps. *Proc Natl Acad Sci U S A*, *88*, 7289-7293.

Guo, Y., Chen, Y., Ito, H., Watanabe, A., Ge, X., Kodama, T. & Aburatani, H. (2006). Identification and characterization of lin-28 homolog B (LIN28B) in human hepatocellular carcinoma. *Gene*, *384*, 51-61.

Haber, M., Burkhart, C. A., Regl, D. L., Madafiglio, J., Norris, M. D. & Horwitz, S. B. (1995). Altered expression of M beta 2, the class II beta-tubulin isotype, in a murine J774.2 cell line with a high level of taxol resistance. *J Biol Chem*, *270*, 31269-31275.

Han, E. K., Gehrke, L., Tahir, S. K., Credo, R. B., Cherian, S. P., Sham, H., Rosenberg, S. H. & Ng, S. (2000). Modulation of drug resistance by alpha-tubulin in paclitaxel-resistant human lung cancer cell lines. *Eur J Cancer*, *36*, 1565-1571.

Hanash, S. M., Kuick, R., Nichols, D. & Stoolman, L. (1988). Quantitative analysis of a new marker for common acute lymphoblastic leukemia detected by two-dimensional electrophoresis. *Dis Markers*, *6*, 209-220.

Henningsson, A., Marsh, S., Loos, W. J., Karlsson, M. O., Garsa, A., Mross, K., Mielke, S., Vigano, L., Locatelli, A., Verweij, J., Sparreboom, A. & McLeod, H. L. (2005). Association of CYP2C8, CYP3A4, CYP3A5, and

ABCB1 polymorphisms with the pharmacokinetics of paclitaxel. *Clin Cancer Res*, *11*, 8097-8104.

Hieber, U. & Heim, M. E. (1994). Tumor necrosis factor for the treatment of malignancies. *Oncology*, *51*, 142-153.

Holmes, F. A., Walters, R. S., Theriault, R. L., Forman, A. D., Newton, L. K., Raber, M. N., Buzdar, A. U., Frye, D. K. & Hortobagyi, G. N. (1991). Phase II trial of taxol, an active drug in the treatment of metastatic breast cancer. *J Natl Cancer Inst*, *83*, 1797-1805.

Holzmayer, T. A., Hilsenbeck, S., Von Hoff, D. D. & Roninson, I. B. (1992). Clinical correlates of MDR1 (P-glycoprotein) gene expression in ovarian and small-cell lung carcinomas. *J Natl Cancer Inst*, *84*, 1486-1491.

Hopper-Borge, E., Chen, Z. S., Shchaveleva, I., Belinsky, M. G. & Kruh, G. D. (2004). Analysis of the drug resistance profile of multidrug resistance protein 7 (ABCC10), resistance to docetaxel. *Cancer Res*, *64*, 4927-4930.

Hopper-Borge, E., Xu, X., Shen, T., Shi, Z., Chen, Z. S. & Kruh, G. D. (2009). Human multidrug resistance protein 7 (ABCC10) is a resistance factor for nucleoside analogues and epothilone B. *Cancer Res*, *69*, 178-184.

Hopper-Borge, E. A., Churchill, T., Paulose, C., Nicolas, E., Jacobs, J. D., Ngo, O., Kuang, Y., Grinberg, A., Westphal, H., Chen, Z. S., Klein-Szanto, A. J., Belinsky, M. G. & Kruh, G. D. (2011). Contribution of Abcc10 (Mrp7) to in vivo paclitaxel resistance as assessed in Abcc10(-/-) mice. *Cancer Res*, *71*, 3649-3657.

Hopper, E., Belinsky, M. G., Zeng, H., Tosolini, A., Testa, J. R. & Kruh, G. D. (2001). Analysis of the structure and expression pattern of MRP7 (ABCC10), a new member of the MRP subfamily. *Cancer Lett*, *162*, 181-191.

Horwitz, S. B. (1994). Taxol (paclitaxel), mechanisms of action. *Ann Oncol*, *5* Suppl 6, S3-6.

Horwitz, S. B., Shen, H. J., He, L., Dittmar, P., Neef, R., Chen, J. & Schubart, U. K. (1997). The microtubule-destabilizing activity of metablastin (p19) is controlled by phosphorylation. *J Biol Chem*, *272*, 8129-8132.

Hu, S., Niu, H., Inaba, H., Orwick, S., Rose, C., Panetta, J. C., Yang, S., Pounds, S., Fan, Y., Calabrese, C., Rehg, J. E., Campana, D., Rubnitz, J. E. & Baker, S. D. (2011). Activity of the multikinase inhibitor sorafenib in combination with cytarabine in acute myeloid leukemia. *J Natl Cancer Inst*, *103*, 893-905.

Huang, Y., Ibrado, A. M., Reed, J. C., Bullock, G., Ray, S., Tang, C. & Bhalla, K. (1997a). Co-expression of several molecular mechanisms of multidrug resistance and their significance for paclitaxel cytotoxicity in human AML HL-60 cells. *Leukemia, 11*, 253-257.

Huang, Y., Ray, S., Reed, J. C., Ibrado, A. M., Tang, C., Nawabi, A. & Bhalla, K. (1997b). Estrogen increases intracellular p26Bcl-2 to p21Bax ratios and inhibits taxol-induced apoptosis of human breast cancer MCF-7 cells. *Breast Cancer Res Treat, 42*, 73-81.

Huisman, M. T., Chhatta, A. A., van Tellingen, O., Beijnen, J. H. & Schinkel, A. H. (2005). MRP2 (ABCC2) transports taxanes and confers paclitaxel resistance and both processes are stimulated by probenecid. *Int J Cancer, 116*, 824-829.

Ibrado, A. M., Liu, L. & Bhalla, K. (1997). Bcl-xL overexpression inhibits progression of molecular events leading to paclitaxel-induced apoptosis of human acute myeloid leukemia HL-60 cells. *Cancer Res, 57*, 1109-1115.

Ishibashi, M., Nakayama, K., Yeasmin, S., Katagiri, A., Iida, K., Nakayama, N., Fukumoto, M. & Miyazaki, K. (2008). A BTB/POZ gene, NAC-1, a tumor recurrence-associated gene, as a potential target for Taxol resistance in ovarian cancer. *Clin Cancer Res, 14*, 3149-3155.

Jaffrezou, J. P., Dumontet, C., Derry, W. B., Duran, G., Chen, G., Tsuchiya, E., Wilson, L., Jordan, M. A. & Sikic, B. I. (1995). Novel mechanism of resistance to paclitaxel (Taxol) in human K562 leukemia cells by combined selection with PSC 833. *Oncol Res, 7*, 517-527.

James, S. W., Silflow, C. D., Stroom, P. & Lefebvre, P. A. (1993). A mutation in the alpha 1-tubulin gene of Chlamydomonas reinhardtii confers resistance to anti-microtubule herbicides. *J Cell Sci, 106* (Pt 1), 209-218.

Johnson, S. M., Grosshans, H., Shingara, J., Byrom, M., Jarvis, R., Cheng, A., Labourier, E., Reinert, K. L., Brown, D. & Slack, F. J. (2005). RAS is regulated by the let-7 microRNA family. *Cell, 120*, 635-647.

Jordan, M. A., Toso, R. J., Thrower, D. & Wilson, L. (1993). Mechanism of mitotic block and inhibition of cell proliferation by taxol at low concentrations. *Proc Natl Acad Sci U S A, 90*, 9552-9556.

Jordan, M. A., Wendell, K., Gardiner, S., Derry, W. B., Copp, H. & Wilson, L. (1996). Mitotic block induced in HeLa cells by low concentrations of paclitaxel (Taxol) results in abnormal mitotic exit and apoptotic cell death. *Cancer Res, 56*, 816-825.

Juliano, R. L. & Ling, V. (1976). A surface glycoprotein modulating drug permeability in Chinese hamster ovary cell mutants. *Biochim Biophys Acta, 455*, 152-162.

Kar, S., Fan, J., Smith, M. J., Goedert, M. & Amos, L. A. (2003). Repeat motifs of tau bind to the insides of microtubules in the absence of taxol. *EMBO J*, *22*, 70-77.

Katsaros, D., Fracchioli, S., Arts, H. J., de Vries, E. G., Danese, S., Richiardi, G., Arisio, R., Gordini, G., Van der Zee, A. G., Suurmeijer, A. J. & Massobrio, M. (1999). [Expression and prognostic value of the drug resistance markers P-gp, Mrp1, Mrp2, and Lrp in ovarian carcinoma]. *Minerva Ginecol*, *51*, 463-470.

Kavallaris, M., Kuo, D. Y., Burkhart, C. A., Regl, D. L., Norris, M. D., Haber, M. & Horwitz, S. B. (1997). Taxol-resistant epithelial ovarian tumors are associated with altered expression of specific beta-tubulin isotypes. *J Clin Invest*, *100*, 1282-1293.

Kelley, M. J., Li, S. & Harpole, D. H. (2001). Genetic analysis of the beta-tubulin gene, TUBB, in non-small-cell lung cancer. *J Natl Cancer Inst*, *93*, 1886-1888.

Khan, I. A. & Luduena, R. F. (1996). Phosphorylation of beta III-tubulin. *Biochemistry*, *35*, 3704-3711.

Kruh, G. D., Guo, Y., Hopper-Borge, E., Belinsky, M. G. & Chen, Z. S. (2007). ABCC10, ABCC11, and ABCC12. *Pflugers Arch*, *453*, 675-684.

Kuang, Y. H., Shen, T., Chen, X., Sodani, K., Hopper-Borge, E., Tiwari, A. K., Lee, J. W., Fu, L. W. & Chen, Z. S. (2010). Lapatinib and erlotinib are potent reversal agents for MRP7 (ABCC10)-mediated multidrug resistance. *Biochem Pharmacol*, *79*, 154-161.

Lamendola, D. E., Duan, Z., Penson, R. T., Oliva, E. & Seiden, M. V. (2003). beta tubulin mutations are rare in human ovarian carcinoma. *Anticancer Res*, *23*, 681-686.

Larcher, J. C., Boucher, D., Lazereg, S., Gros, F. & Denoulet, P. (1996). Interaction of kinesin motor domains with alpha-and beta-tubulin subunits at a tau-independent binding site. Regulation by polyglutamylation. *J Biol Chem*, *271*, 22117-22124.

Larsson, N., Segerman, B., Gradin, H. M., Wandzioch, E., Cassimeris, L. & Gullberg, M. (1999). Mutations of oncoprotein 18/stathmin identify tubulin-directed regulatory activities distinct from tubulin association. *Mol Cell Biol*, *19*, 2242-2250.

Lawrence, C. J., Dawe, R. K., Christie, K. R., Cleveland, D. W., Dawson, S. C., Endow, S. A., Goldstein, L. S., Goodson, H. V., Hirokawa, N., Howard, J., Malmberg, R. L., McIntosh, J. R., Miki, H., Mitchison, T. J., Okada, Y., Reddy, A. S., Saxton, W. M., Schliwa, M., Scholey, J. M., Vale,

R. D., Walczak, C. E. & Wordeman, L. (2004). A standardized kinesin nomenclature. *J Cell Biol*, *167*, 19-22.

Le, X. F., Pruefer, F. & Bast, R. C., Jr. (2005). HER2-targeting antibodies modulate the cyclin-dependent kinase inhibitor p27Kip1 via multiple signaling pathways. *Cell Cycle*, *4*, 87-95.

Lee, K., Klein-Szanto, A. J. & Kruh, G. D. (2000). Analysis of the MRP4 drug resistance profile in transfected NIH3T3 cells. *J Natl Cancer Inst*, *92*, 1934-1940.

Lee, Y. S. & Dutta, A. (2007). The tumor suppressor microRNA let-7 represses the HMGA2 oncogene. *Genes Dev*, *21*, 1025-1030.

Leizer, A. L., Alvero, A. B., Fu, H. H., Holmberg, J. C., Cheng, Y. C., Silasi, D. A., Rutherford, T. & Mor, G. (2011). Regulation of inflammation by the NF-kappaB pathway in ovarian cancer stem cells. *Am J Reprod Immunol*, *65*, 438-447.

Lewis, S. A., Gu, W. & Cowan, N. J. (1987). Free intermingling of mammalian beta-tubulin isotypes among functionally distinct microtubules. *Cell*, *49*, 539-548.

Lewis, S. A., Lee, M. G. & Cowan, N. J. (1985). Five mouse tubulin isotypes and their regulated expression during development. *J Cell Biol*, *101*, 852-861.

Lin, T. L. & Matsui, W. (2012). Hedgehog pathway as a drug target, Smoothened inhibitors in development. *Onco Targets Ther*, *5*, 47-58.

Liptrott, N. J., Pushpakom, S., Wyen, C., Fatkenheuer, G., Hoffmann, C., Mauss, S., Knechten, H., Brockmeyer, N. H., Hopper-Borge, E., Siccardi, M., Back, D. J., Khoo, S. H., Pirmohamed, M. & Owen, A. (2012). Association of ABCC10 polymorphisms with nevirapine plasma concentrations in the German Competence Network for HIV/AIDS. *Pharmacogenet Genomics*, *22*, 10-19.

Litman, T., Brangi, M., Hudson, E., Fetsch, P., Abati, A., Ross, D. D., Miyake, K., Resau, J. H. & Bates, S. E. (2000). The multidrug-resistant phenotype associated with overexpression of the new ABC half-transporter, MXR (ABCG2). *J Cell Sci*, *113,* (Pt 11), 2011-2021.

Liu, F. S. (2009). Mechanisms of chemotherapeutic drug resistance in cancer therapy--a quick review. *Taiwan J Obstet Gynecol*, *48*, 239-244.

Liu, J., Chen, H., Miller, D. S., Saavedra, J. E., Keefer, L. K., Johnson, D. R., Klaassen, C. D. & Waalkes, M. P. (2001). Overexpression of glutathione S-transferase II and multidrug resistance transport proteins is associated with acquired tolerance to inorganic arsenic. *Mol Pharmacol*, *60*, 302-309.

Liu, M., Dai, B., Kang, S. H., Ban, K., Huang, F. J., Lang, F. F., Aldape, K. D., Xie, T. X., Pelloski, C. E., Xie, K., Sawaya, R. & Huang, S. (2006). FoxM1B is overexpressed in human glioblastomas and critically regulates the tumorigenicity of glioma cells. *Cancer Res*, *66*, 3593-3602.

Lu, Q. & Luduena, R. F. (1993). Removal of beta III isotype enhances taxol induced microtubule assembly. *Cell Struct Funct*, *18*, 173-182.

Luo, D., Cheng, S. C., Xie, H. & Xie, Y. (2000). Effects of Bcl-2 and Bcl-XL protein levels on chemoresistance of hepatoblastoma HepG2 cell line. *Biochem Cell Biol*, *78*, 119-126.

Lv, K., Liu, L., Wang, L., Yu, J., Liu, X., Cheng, Y., Dong, M., Teng, R., Wu, L., Fu, P., Deng, W., Hu, W. & Teng, L. (2012). Lin28 mediates paclitaxel resistance by modulating p21, Rb and Let-7a miRNA in breast cancer cells. *PLoS One*, *7*, e40008.

Mahon, F. X., Hayette, S., Lagarde, V., Belloc, F., Turcq, B., Nicolini, F., Belanger, C., Manley, P. W., Leroy, C., Etienne, G., Roche, S. & Pasquet, J. M. (2008). Evidence that resistance to nilotinib may be due to BCR-ABL, Pgp, or Src kinase overexpression. *Cancer Res*, *68*, 9809-9816.

Major, M. L., Lepe, R. & Costa, R. H. (2004). Forkhead box M1B transcriptional activity requires binding of Cdk-cyclin complexes for phosphorylation-dependent recruitment of p300/CBP coactivators. *Mol Cell Biol*, *24*, 2649-2661.

Malofeeva, E. V., Domanitskaya, N., Gudima, M. & Hopper-Borge, E. A. (2012). Modulation of the ATPase and transport activities of broad-acting multidrug resistance factor ABCC10 (MRP7). *Cancer Res*, *72*, 6457-6467.

Maney, T., Hunter, A. W., Wagenbach, M. & Wordeman, L. (1998). Mitotic centromere-associated kinesin is important for anaphase chromosome segregation. *J Cell Biol*, *142*, 787-801.

Marchetti, S., de Vries, N. A., Buckle, T., Bolijn, M. J., van Eijndhoven, M. A., Beijnen, J. H., Mazzanti, R., van Tellingen, O. & Schellens, J. H. (2008). Effect of the ATP-binding cassette drug transporters ABCB1, ABCG2, and ABCC2 on erlotinib hydrochloride (Tarceva) disposition in in vitro and in vivo pharmacokinetic studies employing Bcrp1-/-/ Mdr1a/1b-/-(triple-knockout) and wild-type mice. *Mol Cancer Ther*, *7*, 2280-2287.

Marklund, U., Larsson, N., Gradin, H. M., Brattsand, G. & Gullberg, M. (1996). Oncoprotein 18 is a phosphorylation-responsive regulator of microtubule dynamics. *EMBO J*, *15*, 5290-5298.

Marsh, S., Paul, J., King, C. R., Gifford, G., McLeod, H. L. & Brown, R. (2007). Pharmacogenetic assessment of toxicity and outcome after platinum plus taxane chemotherapy in ovarian cancer, the Scottish Randomised Trial in Ovarian Cancer. *J Clin Oncol*, *25*, 4528-4535.

Martello, L. A., Verdier-Pinard, P., Shen, H. J., He, L., Torres, K., Orr, G. A. & Horwitz, S. B. (2003). Elevated levels of microtubule destabilizing factors in a Taxol-resistant/dependent A549 cell line with an alpha-tubulin mutation. *Cancer Res*, *63*, 1207-1213.

McGuire, W. P., Hoskins, W. J., Brady, M. F., Kucera, P. R., Partridge, E. E., Look, K. Y., Clarke-Pearson, D. L. & Davidson, M. (1996). Cyclophosphamide and cisplatin compared with paclitaxel and cisplatin in patients with stage III and stage IV ovarian cancer. *N Engl J Med*, *334*, 1-6.

Mickley, L. A., Spengler, B. A., Knutsen, T. A., Biedler, J. L. & Fojo, T. (1997). Gene rearrangement, a novel mechanism for MDR-1 gene activation. *J Clin Invest*, *99*, 1947-1957.

Mielke, S. (2007). Individualized pharmacotherapy with paclitaxel. *Curr Opin Oncol*, *19*, 586-589.

Minotti, A. M., Barlow, S. B. & Cabral, F. (1991). Resistance to antimitotic drugs in Chinese hamster ovary cells correlates with changes in the level of polymerized tubulin. *J Biol Chem*, *266*, 3987-3994.

Monzo, M., Rosell, R., Sanchez, J. J., Lee, J. S., O'Brate, A., Gonzalez-Larriba, J. L., Alberola, V., Lorenzo, J. C., Nunez, L., Ro, J. Y. & Martin, C. (1999). Paclitaxel resistance in non-small-cell lung cancer associated with beta-tubulin gene mutations. *J Clin Oncol*, *17*, 1786-1793.

Murphy, W. K., Fossella, F. V., Winn, R. J., Shin, D. M., Hynes, H. E., Gross, H. M., Davilla, E., Leimert, J., Dhingra, H., Raber, M. N. et al. (1993). Phase II study of taxol in patients with untreated advanced non-small-cell lung cancer. *J Natl Cancer Inst*, *85*, 384-388.

Musto, P., Lombardi, G., Matera, R. & Carotenuto, M. (1991). The expression of the multidrug transporter P-170 glycoprotein in remission phase is associated with early and resistant relapse in multiple myeloma. *Haematologica*, *76*, 513-516.

Nakajima, M., Fujiki, Y., Kyo, S., Kanaya, T., Nakamura, M., Maida, Y., Tanaka, M., Inoue, M. & Yokoi, T. (2005). Pharmacokinetics of paclitaxel in ovarian cancer patients and genetic polymorphisms of CYP2C8, CYP3A4, and MDR1. *J Clin Pharmacol*, *45*, 674-682.

Nakamura, T., Furukawa, Y., Nakagawa, H., Tsunoda, T., Ohigashi, H., Murata, K., Ishikawa, O., Ohgaki, K., Kashimura, N., Miyamoto, M., Hirano, S., Kondo, S., Katoh, H., Nakamura, Y. & Katagiri, T. (2004). Genome-wide cDNA microarray analysis of gene expression profiles in pancreatic cancers using populations of tumor cells and normal ductal epithelial cells selected for purity by laser microdissection. *Oncogene*, *23*, 2385-2400.

Nakayama, K., Nakayama, N., Davidson, B., Sheu, J. J., Jinawath, N., Santillan, A., Salani, R., Bristow, R. E., Morin, P. J., Kurman, R. J., Wang, T. L. & Shih Ie, M. (2006). A BTB/POZ protein, NAC-1, is related to tumor recurrence and is essential for tumor growth and survival. *Proc Natl Acad Sci U S A*, *103*, 18739-18744.

Naramoto, H., Uematsu, T., Uchihashi, T., Doto, R., Matsuura, T., Usui, Y., Uematsu, S., Li, X., Takahashi, M., Yamaoka, M. & Furusawa, K. (2007). Multidrug resistance-associated protein 7 expression is involved in cross-resistance to docetaxel in salivary gland adenocarcinoma cell lines. *Int J Oncol*, *30*, 393-401.

Nicoletti, M. I., Valoti, G., Giannakakou, P., Zhan, Z., Kim, J. H., Lucchini, V., Landoni, F., Mayo, J. G., Giavazzi, R. & Fojo, T. (2001). Expression of beta-tubulin isotypes in human ovarian carcinoma xenografts and in a sub-panel of human cancer cell lines from the NCI-Anticancer Drug Screen, correlation with sensitivity to microtubule active agents. *Clin Cancer Res*, *7*, 2912-2922.

Nogales, E., Wolf, S. G. & Downing, K. H. (1998). Structure of the alpha beta tubulin dimer by electron crystallography. *Nature*, *391*, 199-203.

Nylander, K., Marklund, U., Brattsand, G., Gullberg, M. & Roos, G. (1995). Immunohistochemical detection of oncoprotein 18 (Op18) in malignant lymphomas. *Histochem J*, *27*, 155-160.

O'Connor, P. M., Jackman, J., Bae, I., Myers, T. G., Fan, S., Mutoh, M., Scudiero, D. A., Monks, A., Sausville, E. A., Weinstein, J. N., Friend, S., Fornace, A. J. Jr. & Kohn, K. W. (1997). Characterization of the p53 tumor suppressor pathway in cell lines of the National Cancer Institute anticancer drug screen and correlations with the growth-inhibitory potency of 123 anticancer agents. *Cancer Res*, *57*, 4285-4300.

Ogretmen, B. & Safa, A. R. (1996). Down-regulation of apoptosis-related bcl-2 but not bcl-xL or bax proteins in multidrug-resistant MCF-7/Adr human breast cancer cells. *Int J Cancer*, *67*, 608-614.

Oguri, T., Ozasa, H., Uemura, T., Bessho, Y., Miyazaki, M., Maeno, K., Maeda, H., Sato, S. & Ueda, R. (2008). MRP7/ABCC10 expression is a predictive biomarker for the resistance to paclitaxel in non-small cell lung cancer. *Mol Cancer Ther*, *7*, 1150-1155.

Orr, G. A., Verdier-Pinard, P., McDaid, H. & Horwitz, S. B. (2003). Mechanisms of Taxol resistance related to microtubules. *Oncogene*, *22*, 7280-7295.

Pan, L., Gong, Z., Zhong, Z., Dong, Z., Liu, Q., Le, Y. & Guo, J. (2011). Lin-28 reactivation is required for let-7 repression and proliferation in human small cell lung cancer cells. *Mol Cell Biochem*, *355*, 257-263.

Panda, D., Miller, H. P., Banerjee, A., Luduena, R. F. & Wilson, L. (1994). Microtubule dynamics in vitro are regulated by the tubulin isotype composition. *Proc Natl Acad Sci U S A*, *91*, 11358-11362.

Parekh, H., Wiesen, K. & Simpkins, H. (1997). Acquisition of taxol resistance via P-glycoprotein- and non-P-glycoprotein-mediated mechanisms in human ovarian carcinoma cells. *Biochem Pharmacol*, *53*, 461-470.

Peng, X. X., Tiwari, A. K., Wu, H. C. & Chen, Z. S. (2011). Overexpression of P-glycoprotein induces acquired resistance to imatinib in chronic myeloid leukemia cells. *Chin J Cancer*.

Pentheroudakis, G., Kalogeras, K. T., Wirtz, R. M., Grimani, I., Zografos, G., Gogas, H., Stropp, U., Pectasides, D., Skarlos, D., Hennig, G., Samantas, E., Bafaloukos, D., Papakostas, P., Kalofonos, H. P., Pavlidis, N. & Fountzilas, G. (2009). Gene expression of estrogen receptor, progesterone receptor and microtubule-associated protein Tau in high-risk early breast cancer, a quest for molecular predictors of treatment benefit in the context of a Hellenic Cooperative Oncology Group trial. *Breast Cancer Res Treat*, *116*, 131-143.

Perkins, N. D. (2012). The diverse and complex roles of NF-kappaB subunits in cancer. *Nat Rev Cancer*, *12*, 121-132.

Pilarsky, C., Wenzig, M., Specht, T., Saeger, H. D. & Grutzmann, R. (2004). Identification and validation of commonly overexpressed genes in solid tumors by comparison of microarray data. *Neoplasia*, *6*, 744-750.

Pinedo, H. M., Longo, D. L. & Chabner, B. A. (1996). Cancer chemotherapy and biological response modifiers annual 16. Introduction. *Cancer Chemother Biol Response Modif*, *16*, xvii-xviii.

Pirker, R., Wallner, J., Geissler, K., Linkesch, W., Haas, O. A., Bettelheim, P., Hopfner, M., Scherrer, R., Valent, P. & Havelec, L. et al. (1991). MDR1 gene expression and treatment outcome in acute myeloid leukemia. *J Natl Cancer Inst*, *83*, 708-712.

Press, O. W., LeBlanc, M., O'Rourke, T. J., Gagnet, S., Chapman, R. A., Balcerzak, S. P. & Fisher, R. I. (1998). Phase II trial of paclitaxel by 24-hour continuous infusion for relapsed non-Hodgkin's lymphomas, Southwest Oncology Group trial 9246. *J Clin Oncol*, *16*, 574-578.

Pushpakom, S. P., Liptrott, N. J., Rodriguez-Novoa, S., Labarga, P., Soriano, V., Albalater, M., Hopper-Borge, E., Bonora, S., Di Perri, G., Back, D. J., Khoo, S., Pirmohamed, M. & Owen, A. (2011). Genetic variants of ABCC10, a novel tenofovir transporter, are associated with kidney tubular dysfunction. *J Infect Dis*, *204*, 145-153.

Pusztai, L., Jeong, J. H., Gong, Y., Ross, J. S., Kim, C., Paik, S., Rouzier, R., Andre, F., Hortobagyi, G. N., Wolmark, N. & Symmans, W. F. (2009). Evaluation of microtubule-associated protein-Tau expression as a prognostic and predictive marker in the NSABP-B 28 randomized clinical trial. *J Clin Oncol*, *27*, 4287-4292.

Ranganathan, P., Weaver, K. L. & Capobianco, A. J. (2011). Notch signalling in solid tumours, a little bit of everything but not all the time. *Nat Rev Cancer*, *11*, 338-351.

Ranganathan, S., Benetatos, C. A., Colarusso, P. J., Dexter, D. W. & Hudes, G. R. (1998). Altered beta-tubulin isotype expression in paclitaxel-resistant human prostate carcinoma cells. *Br J Cancer*, *77*, 562-566.

Ranganathan, S., McCauley, R. A., Dexter, D. W. & Hudes, G. R. (2001). Modulation of endogenous beta-tubulin isotype expression as a result of human beta(III)cDNA transfection into prostate carcinoma cells. *Br J Cancer*, *85*, 735-740.

Rao, S., Horwitz, S. B. & Ringel, I. (1992). Direct photoaffinity labeling of tubulin with taxol. *J Natl Cancer Inst*, *84*, 785-788.

Rao, S., Krauss, N. E., Heerding, J. M., Swindell, C. S., Ringel, I., Orr, G. A. & Horwitz, S. B. (1994). 3'-(p-azidobenzamido)taxol photolabels the N-terminal 31 amino acids of beta-tubulin. *J Biol Chem*, *269*, 3132-3134.

Rao, S., Orr, G. A., Chaudhary, A. G., Kingston, D. G. & Horwitz, S. B. (1995). Characterization of the taxol binding site on the microtubule. 2-(m-Azidobenzoyl)taxol photolabels a peptide (amino acids 217-231) of beta-tubulin. *J Biol Chem*, *270*, 20235-20238.

Regnard, C., Desbruyeres, E., Denoulet, P. & Edde, B. (1999). Tubulin polyglutamylase, isozymic variants and regulation during the cell cycle in HeLa cells. *J Cell Sci*, *112* (Pt 23), 4281-4289.

Reiss, M., Brash, D. E., Munoz-Antonia, T., Simon, J. A., Ziegler, A., Vellucci, V. F. & Zhou, Z. L. (1992). Status of the p53 tumor suppressor gene in human squamous carcinoma cell lines. *Oncol Res*, *4*, 349-357.

Reya, T., Duncan, A. W., Ailles, L., Domen, J., Scherer, D. C., Willert, K., Hintz, L., Nusse, R. & Weissman, I. L. (2003). A role for Wnt signalling in self-renewal of haematopoietic stem cells. *Nature*, *423*, 409-414.

Robert, J. (1999). Multidrug resistance in oncology, diagnostic and therapeutic approaches. *Eur J Clin Invest*, *29*, 536-545.

Rose, W. C. (1993). Taxol-based combination chemotherapy and other in vivo preclinical antitumor studies. *J Natl Cancer Inst Monogr*, 47-53.

Rosell, R., Gonzalez-Larriba, J. L., Alberola, V., Molina, F., Monzo, M., Benito, D., Perez, J. M. & de Anta, J. M. (1995). Single-agent paclitaxel by 3-hour infusion in the treatment of non-small cell lung cancer, links between p53 and K-ras gene status and chemosensitivity. *Semin Oncol*, *22*, 12-18.

Rudin, D., Li, L., Niu, N., Kalari, K. R., Gilbert, J. A., Ames, M. M. & Wang, L. (2011). Gemcitabine Cytotoxicity, Interaction of Efflux and Deamination. *J Drug Metab Toxicol*, *2*, 1-10.

Rudner, A. D. & Murray, A. W. (1996). The spindle assembly checkpoint. *Curr Opin Cell Biol*, *8*, 773-780.

Sarkadi, B., Homolya, L., Szakacs, G. & Varadi, A. (2006). Human multidrug resistance ABCB and ABCG transporters, participation in a chemoimmunity defense system. *Physiol Rev*, *86*, 1179-1236.

Sato, H., Gottesman, M. M., Goldstein, L. J., Pastan, I., Block, A. M., Sandberg, A. A. & Preisler, H. D. (1990). Expression of the multidrug resistance gene in myeloid leukemias. *Leuk Res*, *14*, 11-21.

Sauna, Z. E., Smith, M. M., Muller, M., Kerr, K. M. & Ambudkar, S. V. (2001). The mechanism of action of multidrug-resistance-linked P-glycoprotein. *J Bioenerg Biomembr*, *33*, 481-491.

Saville, M. W., Lietzau, J., Pluda, J. M., Feuerstein, I., Odom, J., Wilson, W. H., Humphrey, R. W., Feigal, E., Steinberg, S. M. & Broder, S. et al. (1995). Treatment of HIV-associated Kaposi's sarcoma with paclitaxel. *Lancet*, *346*, 26-28.

Schiff, P. B., Fant, J. & Horwitz, S. B. (1979). Promotion of microtubule assembly in vitro by taxol. *Nature*, *277*, 665-667.

Schinkel, A. H. & Jonker, J. W. (2003). Mammalian drug efflux transporters of the ATP binding cassette (ABC) family, an overview. *Adv Drug Deliv Rev*, *55*, 3-29.

Schinkel, A. H., Mayer, U., Wagenaar, E., Mol, C. A., van Deemter, L., Smit, J. J., van der Valk, M. A., Voordouw, A. C., Spits, H., van Tellingen, O., Zijlmans, J. M., Fibbe, W. E. & Borst, P. (1997). Normal viability and altered pharmacokinetics in mice lacking mdr1-type (drug-transporting) P-glycoproteins. *Proc Natl Acad Sci U S A*, *94*, 4028-4033.

Schondorf, T., Kurbacher, C. M., Gohring, U. J., Benz, C., Becker, M., Sartorius, J., Kolhagen, H., Mallman, P. & Neumann, R. (2002). Induction of MDR1-gene expression by antineoplastic agents in ovarian cancer cell lines. *Anticancer Res*, *22*, 2199-2203.

Schondorf, T., Scharl, A., Kurbacher, C. M., Bien, O., Becker, M., Neumann, R., Kolhagen, H., Rustemeyer, J., Mallmann, P. & Gohring, U. J. (1999). Amplification of the mdr1-gene is uncommon in recurrent ovarian carcinomas. *Cancer Lett*, *146*, 195-199.

Shao, Y. Y., Kuo, K. T., Hu, F. C., Lu, Y. S., Huang, C. S., Liau, J. Y., Lee, W. C., Hsu, C., Kuo, W. H., Chang, K. J., Lin, C. H. & Cheng, A. L. (2010). Predictive and prognostic values of tau and ERCC1 in advanced breast cancer patients treated with paclitaxel and cisplatin. *Jpn J Clin Oncol*, *40*, 286-293.

Sharma, N., Ramachandran, S., Bowers, M., Yegappan, M., Brown, R., Aziz, S., Chapman, R. & Yu, B. W. (2000). Multiple factors other than p53 influence colon cancer sensitivity to paclitaxel. *Cancer Chemother Pharmacol*, *46*, 329-337.

Shen, T., Kuang, Y. H., Ashby, C. R., Lei, Y., Chen, A., Zhou, Y., Chen, X., Tiwari, A. K., Hopper-Borge, E., Ouyang, J. & Chen, Z. S. (2009). Imatinib and nilotinib reverse multidrug resistance in cancer cells by inhibiting the efflux activity of the MRP7 (ABCC10). *PLoS One*, *4*, e7520.

Siegel, R., Naishadham, D. & Jemal, A. (2013). Cancer statistics, 2013. *CA Cancer J Clin*, *63*, 11-30.

Singh, S., Chitkara, D., Mehrazin, R., Behrman, S. W., Wake, R. W. & Mahato, R. I. (2012). Chemoresistance in prostate cancer cells is regulated by miRNAs and Hedgehog pathway. *PLoS One*, *7*, e40021.

Sissung, T. M., Mross, K., Steinberg, S. M., Behringer, D., Figg, W. D., Sparreboom, A. & Mielke, S. (2006). Association of ABCB1 genotypes with paclitaxel-mediated peripheral neuropathy and neutropenia. *Eur J Cancer*, *42*, 2893-2896.

Smith, A. J., van Helvoort, A., van Meer, G., Szabo, K., Welker, E., Szakacs, G., Varadi, A., Sarkadi, B. & Borst, P. (2000). MDR3 P-glycoprotein, a phosphatidylcholine translocase, transports several cytotoxic drugs and directly interacts with drugs as judged by interference with nucleotide trapping. *J Biol Chem*, *275*, 23530-23539.

Snyder, J. P., Nettles, J. H., Cornett, B., Downing, K. H. & Nogales, E. (2001). The binding conformation of Taxol in beta-tubulin, a model based on electron crystallographic density. *Proc Natl Acad Sci U S A*, *98*, 5312-5316.

Sodani, K., Patel, A., Kathawala, R. J. & Chen, Z. S. (2012). Multidrug resistance associated proteins in multidrug resistance. *Chin J Cancer*, *31*, 58-72.

Song, Z., Yue, W., Wei, B., Wang, N., Li, T., Guan, L., Shi, S., Zeng, Q., Pei, X. & Chen, L. (2011). Sonic hedgehog pathway is essential for maintenance of cancer stem-like cells in human gastric cancer. *PLoS One*, *6*, e17687.

Sparreboom, A., van Asperen, J., Mayer, U., Schinkel, A. H., Smit, J. W., Meijer, D. K., Borst, P., Nooijen, W. J., Beijnen, J. H. & van Tellingen, O. (1997). Limited oral bioavailability and active epithelial excretion of paclitaxel (Taxol) caused by P-glycoprotein in the intestine. *Proc Natl Acad Sci U S A*, *94*, 2031-2035.

Srivastava, R. K., Srivastava, A. R., Korsmeyer, S. J., Nesterova, M., Cho-Chung, Y. S. & Longo, D. L. (1998). Involvement of microtubules in the regulation of Bcl2 phosphorylation and apoptosis through cyclic AMP-dependent protein kinase. *Mol Cell Biol*, *18*, 3509-3517.

Stearns, M. E. & Wang, M. (1992). Taxol blocks processes essential for prostate tumor cell (PC-3 ML) invasion and metastases. *Cancer Res*, *52*, 3776-3781.

Steg, A. D., Bevis, K. S., Katre, A. A., Ziebarth, A., Dobbin, Z. C., Alvarez, R. D., Zhang, K., Conner, M. & Landen, C. N. (2012). Stem cell pathways contribute to clinical chemoresistance in ovarian cancer. *Clin Cancer Res*, *18*, 869-881.

Strobel, T., Swanson, L., Korsmeyer, S. & Cannistra, S. A. (1996). BAX enhances paclitaxel-induced apoptosis through a p53-independent pathway. *Proc Natl Acad Sci U S A*, *93*, 14094-14099.

Sun, Y. L., Chen, J. J., Kumar, P., Chen, K., Sodani, K., Patel, A., Chen, Y. L., Chen, S. D., Jiang, W. Q. & Chen, Z. S. (2013). Reversal of MRP7 (ABCC10)-Mediated Multidrug Resistance by Tariquidar. *PLoS One*, *8*, e55576.

Takayanagi, S., Kataoka, T., Ohara, O., Oishi, M., Kuo, M. T. & Ishikawa, T. (2004). Human ATP-binding cassette transporter ABCC10, expression profile and p53-dependent upregulation. *J Exp Ther Oncol*, *4*, 239-246.

Tiwari, A. K., Sodani, K., Dai, C. L., Abuznait, A. H., Singh, S., Xiao, Z. J., Patel, A., Talele, T. T., Fu, L., Kaddoumi, A., Gallo, J. M. & Chen, Z. S. (2013). Nilotinib potentiates anticancer drug sensitivity in murine ABCB1-, ABCG2-, and ABCC10-multidrug resistance xenograft models. *Cancer Lett*, *328*, 307-317.

Tiwari, A. K., Sodani, K., Dai, C. L., Ashby, C. R., Jr. & Chen, Z. S. (2011). Revisiting the ABCs of multidrug resistance in cancer chemotherapy. *Curr Pharm Biotechnol*, *12*, 570-594.

Ueda, K., Cornwell, M. M., Gottesman, M. M., Pastan, I., Roninson, I. B., Ling, V. & Riordan, J. R. (1986). The mdr1 gene, responsible for multidrug-resistance, codes for P-glycoprotein. *Biochem Biophys Res Commun*, *141*, 956-962.

Ulasov, I. V., Nandi, S., Dey, M., Sonabend, A. M. & Lesniak, M. S. (2011). Inhibition of Sonic hedgehog and Notch pathways enhances sensitivity of CD133(+) glioma stem cells to temozolomide therapy. *Mol Med*, *17*, 103-112.

van der Zee, A. G., Hollema, H., Suurmeijer, A. J., Krans, M., Sluiter, W. J., Willemse, P. H., Aalders, J. G. & de Vries, E. G. (1995). Value of P-glycoprotein, glutathione S-transferase pi, c-erbB-2, and p53 as prognostic factors in ovarian carcinomas. *J Clin Oncol*, *13*, 70-78.

Vasey, P. A., Jones, N. A., Jenkins, S., Dive, C. & Brown, R. (1996). Cisplatin, camptothecin, and taxol sensitivities of cells with p53-associated multidrug resistance. *Mol Pharmacol*, *50*, 1536-1540.

Verrelle, P., Meissonnier, F., Fonck, Y., Feillel, V., Dionet, C., Kwiatkowski, F., Plagne, R. & Chassagne, J. (1991). Clinical relevance of immunohistochemical detection of multidrug resistance P-glycoprotein in breast carcinoma. *J Natl Cancer Inst*, *83*, 111-116.

Walle, T., Walle, U. K., Kumar, G. N. & Bhalla, K. N. (1995). Taxol metabolism and disposition in cancer patients. *Drug Metab Dispos*, *23*, 506-512.

Wang, J., Rao, S., Chu, J., Shen, X., Levasseur, D. N., Theunissen, T. W. & Orkin, S. H. (2006). A protein interaction network for pluripotency of embryonic stem cells. *Nature*, *444*, 364-368.

Wang, P., Zhang, Z., Gao, K., Deng, Y., Zhao, J., Liu, B. & Li, X. (2009). [Expression and Clinical Significance of ABCC10 in the Patients with Non-small Cell Lung Cancer.]. *Zhongguo Fei Ai Za Zhi*, *12*, 875-878.

Wang, T. H., Wang, H. S. & Soong, Y. K. (2000). Paclitaxel-induced cell death, where the cell cycle and apoptosis come together. *Cancer*, *88*, 2619-2628.

Wani, M. C., Taylor, H. L., Wall, M. E., Coggon, P. & McPhail, A. T. (1971). Plant antitumor agents. VI. The isolation and structure of taxol, a novel antileukemic and antitumor agent from Taxus brevifolia. *J Am Chem Soc*, *93*, 2325-2327.

Whitfield, C., Abraham, I., Ascherman, D. & Gottesman, M. M. (1986). Transfer and amplification of a mutant beta-tubulin gene results in colcemid dependence, use of the transformant to demonstrate regulation of beta-tubulin subunit levels by protein degradation. *Mol Cell Biol*, *6*, 1422-1429.

Wilson, L. & Jordan, M. A. (1995). Microtubule dynamics, taking aim at a moving target. *Chem Biol*, *2*, 569-573.

Wilson, W. H., Chabner, B. A., Bryant, G., Bates, S., Fojo, A., Regis, J., Jaffe, E. S., Steinberg, S. M., Goldspiel, B. R. & Cheson, B. D. et al. (1995). Phase II study of paclitaxel in relapsed non-Hodgkin's lymphomas. *J Clin Oncol*, *13*, 381-386.

Wu, C. P., Calcagno, A. M. & Ambudkar, S. V. (2008). Reversal of ABC drug transporter-mediated multidrug resistance in cancer cells, evaluation of current strategies. *Curr Mol Pharmacol*, *1*, 93-105.

Xu, B., Zhang, K. & Huang, Y. (2009). Lin28 modulates cell growth and associates with a subset of cell cycle regulator mRNAs in mouse embryonic stem cells. *RNA*, *15*, 357-361.

Xu, X., Leo, C., Jang, Y., Chan, E., Padilla, D., Huang, B. C., Lin, T., Gururaja, T., Hitoshi, Y., Lorens, J. B., Anderson, D. C., Sikic, B., Luo, Y., Payan, D. G. & Nolan, G. P. (2001). Dominant effector genetics in mammalian cells. *Nat Genet*, *27*, 23-29.

Yang, X., Lin, X., Zhong, X., Kaur, S., Li, N., Liang, S., Lassus, H., Wang, L., Katsaros, D., Montone, K., Zhao, X., Zhang, Y., Butzow, R., Coukos, G. & Zhang, L. (2010). Double-negative feedback loop between reprogramming factor LIN28 and microRNA let-7 regulates aldehyde dehydrogenase 1-positive cancer stem cells. *Cancer Res*, *70*, 9463-9472.

Yao, J., An, Y., Wie, J. S., Ji, Z. L., Lu, Z. P., Wu, J. L., Jiang, K. R., Chen, P., Xu, Z. K. & Miao, Y. (2011). Cyclopamine reverts acquired chemoresistance and down-regulates cancer stem cell markers in pancreatic cancer cell lines. *Swiss Med Wkly*, *141*, w13208.

Yin, S., Bhattacharya, R. & Cabral, F. (2010). Human mutations that confer paclitaxel resistance. *Mol Cancer Ther*, *9*, 327-335.

Yip, N. C., Fombon, I. S., Liu, P., Brown, S., Kannappan, V., Armesilla, A. L., Xu, B., Cassidy, J., Darling, J. L. & Wang, W. (2011). Disulfiram modulated ROS-MAPK and NFkappaB pathways and targeted breast cancer cells with cancer stem cell-like properties. *Br J Cancer*, *104*, 1564-1574.

Yusuf, R. Z., Duan, Z., Lamendola, D. E., Penson, R. T. & Seiden, M. V. (2003). Paclitaxel resistance, molecular mechanisms and pharmacologic manipulation. *Curr Cancer Drug Targets*, *3*, 1-19.

Zaffaroni, N., Silvestrini, R., Orlandi, L., Bearzatto, A., Gornati, D. & Villa, R. (1998). Induction of apoptosis by taxol and cisplatin and effect on cell cycle-related proteins in cisplatin-sensitive and -resistant human ovarian cells. *Br J Cancer*, *77*, 1378-1385.

Zaman, G. J., Flens, M. J., van Leusden, M. R., de Haas, M., Mulder, H. S., Lankelma, J., Pinedo, H. M., Scheper, R. J., Baas, F. & Broxterman, H. J. et al. (1994). The human multidrug resistance-associated protein MRP is a plasma membrane drug-efflux pump. *Proc Natl Acad Sci U S A*, *91*, 8822-8826.

Zhang, C. C., Yang, J. M., Bash-Babula, J., White, E., Murphy, M., Levine, A. J. & Hait, W. N. (1999). DNA damage increases sensitivity to vinca alkaloids and decreases sensitivity to taxanes through p53-dependent repression of microtubule-associated protein 4. *Cancer Res*, *59*, 3663-3670.

Zhao, C., Blum, J., Chen, A., Kwon, H. Y., Jung, S. H., Cook, J. M., Lagoo, A. & Reya, T. (2007). Loss of beta-catenin impairs the renewal of normal and CML stem cells in vivo. *Cancer Cell*, *12*, 528-541.

Zhou, J., Cheng, S. C., Luo, D. & Xie, Y. (2001). Study of multi-drug resistant mechanisms in a taxol-resistant hepatocellular carcinoma QGY-TR 50 cell line. *Biochem Biophys Res Commun*, *280*, 1237-1242.

Zhou, Y., Hopper-Borge, E., Shen, T., Huang, X. C., Shi, Z., Kuang, Y. H., Furukawa, T., Akiyama, S., Peng, X. X., Ashby, C. R. Jr., Chen, X., Kruh, G. D. & Chen, Z. S. (2009). Cepharanthine is a potent reversal agent for MRP7(ABCC10)-mediated multidrug resistance. *Biochem Pharmacol*, *77*, 993-1001.

In: Paclitaxel
Editor: Diego Morales

ISBN: 978-1-62808-549-5

Chapter 3

Biosynthesis of Paclitaxel and its Pharmaceutical Application for Cancer Treatment

Vesna D. Nikolic**, *Ivan M. Savic, Ljubisa B. Nikolic and Ivana M. Savic
Faculty of Technology, University of Nis, Leskovac, Serbia

Abstract

Pacific yew (*Taxus brevifolia*) is a common tree of in the coniferous forest understory of western North America. All parts of the plant are poisonous except the red layer of fruits (*arilus*). The last thirty years, the pharmaceutical company are interested in this plant due to the presence of the bioactive compounds such as taxine alkaloids (Taxine A and B) and paclitaxel (Taxol). The chemical name for paclitaxel is 5β,20-Epoxy-1,2α,4,7β,10β,13α-hexahydroxytax-11-en-9-one 4,10-diacetate 2-benzoate 13-ester with (2R,3S)-N-benzoyl-3-phenylisoserine.The concentration of paclitaxel is 0.01-0.02% in the bark of this yew. The structure of paclitaxel is very complex. From this reason and the complicated procedures for the synthesis, the researchers rather isolate this compound from the bark of yew. The pathway of paclitaxel

* Corresponding author: Dr Vesna D. Nikolic, Department of Pharmaceutics, Faculty of Technology, University of Nis, Bulevar oslobodjenja 124, 16000 Leskovac, Serbia. Tel.: +381 16 242859; fax: +381 16 242859. *E-mail address:* ici_teh@yahoo.com.

biosynthesis has 20 steps. The main precursor for biosynthesis is geranylgeranyl diphosphate, which is cyclizated to taxa-4(5),11(12)-diene. The obtained olefin is oxidized to taxa-4(20),11(12)-dien-5α-ol by the cytochrome P450, and then the alcohol is converted to the corresponding acetate ester by the acetyl CoA. After confirmation of the anticancer activity of paclitaxel, the World Health Organisation allowed its application in the treatment of cancer. Paclitaxel simulates the assembly of tubulin dimmers in microtubules. It stabilizes the microtubules and prevents their depolymerization. This induced stability inhibits the normal dynamic reorganization of the microtubules network. It blocks cells in the G2/M phase of the cell cycle and such cells are unable to form a normal mitotic apparatus. The clinical studies were shown that the paclitaxel is effective in the cancer therapy of ovarian, intestine, lung and breast. Today, a few commercial drugs with the paclitaxel as a pharmaceutical active compound are available in the market. The preparations are commonly administered intravenously. It is supplied as a non aqueous solution intended for dilution with a suitable parenteral fluid prior to intravenous infusion. Based on the successful application in the cancer treatment, the paclitaxel has the significant place among the other chemotherapeutic agents even today.

Keywords: Paclitaxel, biosynthesis, clinical study, pharmacological activity

1. Introduction

Nowadays, there are many medical ways, such as surgery, radiotherapy, chemotherapy and immunotherapy for the cancer treatment. Chemotherapy presents a relatively new cancer treatment method [1]. Chemotherapeutic agents interfere with the process by which cancer cells divide to produce new cells. The drugs are introduced into the bloodstream and circulate around the body. These substances kill the cancer cells that reside at the original site of occurrence and those migrated to other tissues (metastasis) [2]. The chemotherapy is limited as the other cancer therapeutic methods. These include the inablity to destroy all cancer cells of a large tumor, the drug resistance, as well as its side effects, such as nausea, vomiting, diarrhea, hair loss and increased infections. The most of these side effects are the result of the normal cells destruction. The natural products become an extremely important during searching for more effective anticancer agents. The treatment of human diseases for thousands of years is possible using the terrestrial plants, although the actual rationale of the plant's effect may have been largely semiempirical or even unknown. Currently, on the market there are over 80 anticancer drugs available. They are especially derived directly from natural sources where the plants is commonly used. They are especially derived directly from natural sources, usually the plants [3]. The one of these efficient agents is paclitaxel.

Paclitaxel is the best known a natural drug against tumor. Botanist Arthur S. Barclay (1962) collected the samples of European yew (*Taxus baccata*), because he was inspired by the historical story about the Gallic chieftain *Cativolcus,* who empoison after drinkig the tea of the leaves of this plant. He doubted that the plant contains the cytotoxic compounds. The crude extracts of the bark of yew, analyzed in the Research Triangle Institute (Carolina, USA), were showed unusual wide spectra of cytotoxic activity against the cells of leukemia and numerous tumor cells. Wall and Wani (1971) isolated the paclitaxel from this samples and completely crystallographically defined the structure.

In the period from 1960 to 1981, paclitaxel was shown the best activity than other 110,000 famous compounds from 35,000 plants. However, its application was not immediately started, because poor solubility was a big problem and very expensive isolation process (in order to obtain 25 kg of paclitaxel, it is necessary about 38,000 trees of *Taxus brevifolia*, what is enough for therapy only 12,000 patients) [4]. Schiff et al., [5] published that the paclitaxel has a unique mechanism of action in the tumor cells, which is significantly different than other antitumor drugs. This has led to renewed interest in the paclitaxel and new research.

The ecosystems were changed by massive cutting *Taxus brevifolia* forests. Nicolau (The Scripps Research Institute, La Jolla) and Holton (Florida State University, Tallahassee) developed the total synthesis of paclitaxel in 1994, because they were motivated by this problem. Despite the fact that there are a six developed total synthesis, paciltaxel is produced by the semisynthetic pathway due to the economic unprofitability of these procedures. Pierre Potier (Institut de Chimie des Substances Naturelles, CNRS, France) discovered the structural similar compound (10-deacetylbaccatin III) in the leaves of European yew (*Taxus baccata*) [6]. This compound can be transformed in the paclitaxel through a few simple steps. This discovery has enabled not only commercial production of paclitaxel, but also the development of new drug, Taxotere® that do not exist in the nature and which is more active than paclitaxel.

Paclitaxel (tax-11-en-9-one-5β,20-epoxyl-1,2α,4,7β,13α-hexahydroxy-4,10-diacetate-2-benzoate 13-β-benzoylamino-α-hydroxybenzenepropionate), a natural diterpenoid pseudoalkaloid, is white to off-white crystalline powder, highly lipophilic [7], insoluble in water and melts at around 216–217 °C (Fig. 1). It is also well known as taxol [8]. Structure determination of paclitaxel was achieved through a combination of spectroscopic techniques, chemical degradation, and X-ray crystallography of derivatives of degraded moieties. It was evident that paclitaxel had the empiric formula of $C_{47}H_{51}NO_{14}$ with a molecular weight of 853.9 Da, and it was comprised of a taxane nucleus to which a rare fourmembered oxetane ring was linked to C-4 and C-5, and an ester was attached at the C-13 position.

This compound is contained in the bark of Pacific yew (*Taxus brevifolia*). It is a small, slow-growing evergreen tree sparsely dispersed along the coastal and interior mountains of the Pacific Northwest [9].

Unlike the Vinca alkaloids, colchicine, and other antimitotic agents that act by preventing the polymerization of tubulin into microtubules, paclitaxel promotes microtubule assembly and suppresses depolymerization [10,11].

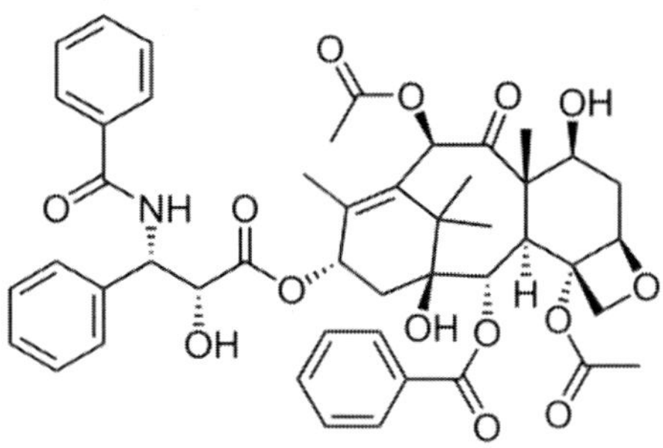

Figure 1. The structure of paclitaxel.

1.1. Biosynthesis of Paclitaxel

Paclitaxel presents a polyoxygenated and highly substituted cyclic diterpenoid [6]. The structure of paclitaxel contains the taxane ring system. It differs than other taxanes either in substitution pattern, the nature of the ester side chains or in the presence of the oxetane ring (D-ring) system. A biosynthetic pathway consists of approximately 20 enzymatic steps [12]. The first step of the paclitaxel biosynthesis is the cyclization of the universal diterpenoid precursor geranylgeranyl diphosphate to taxa-4(5),11(12)-diene (Fig. 2) [13].

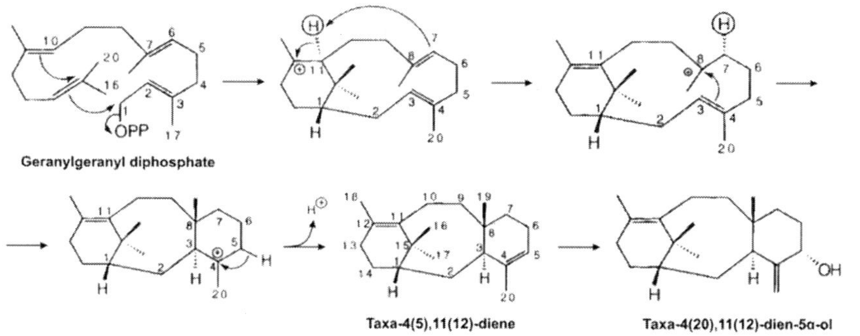

Figure 2. Mechanism for cyclization of geranylgeranyl diphosphate to taxa-4(5),11(12)-diene.

Also, this process is not rate-limiting step, catalyzed by enzyme taxadiene synthase, which prefers Mg^{2+} (K_m - 0.16 mM) to Mn^{2+} and other divalent metal ions as the required cofactor. The enzyme is a 79-kDa diterpene cyclase [14,15] with the properties typical for this class of enzymes. This slow process rather performs than the cyclization to the 4(20),11(12)-diene isomer previously suggested based on the taxoids with double bonds in these position. Actually, Lin et al. [16] explained the mechanism of taxadiene synthase using substrates [1-$^2H_2$, 20-$^2H_3$], [20-$^2H_3$] and [4-$^2H_2$]geranylgeranyl diphosphates with the partially purified taxadiene synthase from Pacific yew steams. The studies indicated that all hydrogens at the C3-methyl group of the acyclic precursor (C20 of geranylgeranyl diphosphate; C20 of taxadiene) are retained in the cyclization whereas one hydrogen atom from the C4-methylene group of geranylgeranyl diphosphate (C5 of taxadiene) is lost. The mechanism of cyclization geranylgeranyl diphosphate to taxa-4(5),11(12)-diene was defined with a minimum number of steps and a single, enzyme-bound, olefinic intermediate. The C-C bond is formed between C1 and C14 of the substrate after ionization of the geranylgeranyl diphosphate ester (1), as well as the A ring is closed via re-face attack at C10. The obtained 11R-verticillyl cation (3) is deprotonized to the 1S-verticillene (4) by removal of the 11R-hydrogen. The taxenyl cation (5) is obtained after the transannular cyclization, catalyzed by the same enzyme. During this process the intermediate (4) is reprotonated at C7 to the intermediate (5). This intermediate is further deprotonated at C5 to generation of the endocyclic double bond of the taxadiene product (6).

In the second biosynthetic step, the taxa-4(5),11(12)-diene is hydroxylated at C-5 position by a cytochrome P450 enzyme [17] with migration of the double bond [18] (Fig. 2). This is the first of eight oxygenation steps (of the taxane core) on route to paclitaxel. The NADPH and O_2 dependent conversion of taxa-4(5),11(12)-diene to taxa-4(20),11(12)-dien-5α-ol [17] is catalyzed by microsomal preparations from Taxus stem [19] and cultured cells. Many terpenoid hydroxylases are commonly the cytochrome P450 type [20]. The mechanism of this step is not completely explained, so it is not yet known whether an epoxide intermediate or direct oxygen insertion at C-5 is involved. The structure of the first oxygenated intermediate indicates the unusual double bond migration that limits the mechanistic possibilities for further elaboration at this site of the oxetane moiety of paclitaxel. The role of alcohol as an intermediate was confirmed by *in vivo* conversion into advanced taxoids, such as 10-deacetylbaccatin III, cephalomannine, and paclitaxel in *T. brevifolia* stem disks [17]. This first oxygenation step is slow in compared with subsequent metabolic transformations, so that suitable for gene isolation.

The third step in the biosynthesis presents the acylation of taxadienol. The acylation of taxa-4(20),11(12)-dien-5α-ol is necessary, because the natural taxoids are acetyled or acylated at the C-5 position [21]. The elaboration of the oxetane ring includes the epoxidation of the 4(20)-double bond and intramolecular acetoxy migration with oxirane ring expansion (Fig. 3) [17,22,23]. The 4(20)-ene-5α-acetate functional group is very important in this process. These considerations and subsequent oxygenation steps led to the evaluation of the enzymatic acylation of taxa-4(20),11(12)-dien-5α-ol. The soluble transacetylase that catalyzes acetyl coenzyme A dependent acylation has been partially purified and characterized as a 50 kDa monomeric protein from *T. canadensis* suspension cultured cells. It accepts acetyl CoA, but not benzoyl CoA as a cosubstrate.

The other reactions on the paclitaxel pathway involve additional oxygenations, hydroxy group acylations, oxidation to a ketone, and generation of the oxetane ring system. Based on the consideration, the order of oxygenation to be C-5 followed by C-10, then C-2 and C-9 [23,24].

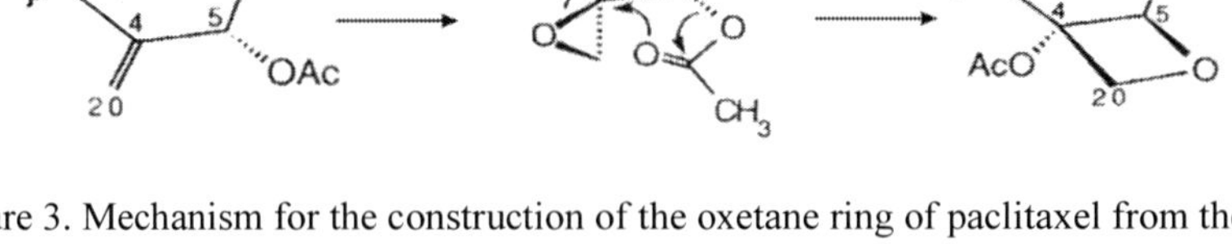

Figure 3. Mechanism for the construction of the oxetane ring of paclitaxel from the 4(20)-ene-5α-acetoxy functional grouping.

1.2. Mechanism of Action

Microtubules are the third principal component of the cytoskeleton of eukaryotic cells [5]. They are important during the cell movements, including some forms of cell locomotion, the intracellular transport of organelles, and the separation of chromosomes during mitosis, as well as to determine cell shape. Microtubules are formed after the polymerization of tubulin. Tubulin is a single type of globular protein and presents a heterodimer that is composed of two polypeptides (α-and β-tubulin). Polymerization of the heterodimeric subunits is regulated by a few microtubule-associated proteins [25]. This process involves the hydrolysis of guanosine 5'-triphosphate. Dynamic instability results in the continual and rapid turnover of the most microtubules, which have half-lives of only several minutes within the cell. This rapid turnover of microtubules is particularly critical for the remodeling of the cytoskeleton that occurs during mitosis. The interactions of microtubules with other proteins can modify the dinamic behavior. Some cellular proteins act to disassemble microtubules, either by severing microtubules or by increasing the rate of tubulin depolymerization from microtubule ends. The microtubule-associated proteins increase their stability after binding for microtubules. Such interactions allow the cell to stabilize microtubules in particular locations and provide an important mechanism for determining cell shape and polarity.

Because of the central role of microtubules in mitosis, compounds that target microtubules are potent cytotoxic agents. One of them is paclitaxel. The cytotoxic activities of this compound have been linked to the stabilization of microtubules [25]. Tubulin is the primary target responsible for the cytotoxic properties of paclitaxel, as reflected by its ability to arrest cells in mitosis [3] and to induce the assembly of tubulin into microtubules *in vitro* [5,27]. Paclitaxel induces microtubule assembly by shifting the equilibrium between tubulin dimers (α-and β -tubulin) and microtubules by lowering the critical concentration of tubulin required for polymerization. After entrance into mitochondria (MT), paclitaxel binds the polymeric MT form of tubulin in a 1:1 stoichiometry with the α-and β -tubulin heterodimer subunits [28]. This leads to hence augmenting and stabilizing the MT pool [29]. *In vitro* paclitaxel has also been shown to nucleate tubulin polymerization into MTs, eliminating the requirement for GTP in normal polymerization [5,30]. Within cells this effect is manifested by micromolar paclitaxel levels overriding the centrosomal microtubule-organizing center function and inducing the appearance of many short non-centrosomally linked bundles throughout the cytoplasm [31]. In rapidly cycling cultured human cells, paclitaxel induces a block at the transition between G and M phase [27].

The numerous investigations were focused for finding the low concentrations of paclitaxel that inhibit mitosis in cancer cells. Long et al. [32] have reported that very low cytotoxic concentrations (3 and 10 nM) of paclitaxel are capable of inducing the formation of multiple micronuclei in HCT116 human colon carcinoma cells as a result of the inhibition of mitotic spindle assembly or function without causing mitotic arrest or microtubule bundling [33]. It is known that paclitaxel has ability to inhibit HeLa cell proliferation by inducing a sustained mitotic block at the metaphase/anaphase boundary. During the study, Jordan et al. [34] founded that the half-maximal inhibition of cell proliferation was occurred at 8 nM paclitaxel, and mitosis was half-maximally blocked at 8 nM paclitaxel. Inhibition of mitosis was associated with formation of an incomplete metaphase plate of chromosomes and an altered arrangement of spindle microtubules. No increase in microtubule polymer mass occurred below 10 nM paclitaxel. The mass of microtubules increased half-maximally at 80 nM paclitaxel and attained maximal levels 330 nM paclitaxel. At submicromolar concentrations, paclitaxel suppressed growing and shortening at the ends of microtubules reassembled *in vitro* from bovine brain tubulin. At its lowest effective concentrations, paclitaxel appears to block mitosis by kinetically stabilizing spindle microtubules and not by changing the mass of polymerized microtubules.

Although treatment of various tumor cells with paclitaxel *in vivo* and *in vitro* induces apoptosis [35], and paclitaxel has been shown to regulate the expression of several apoptosis-related proteins, including bcl-2, bax, bcl-X, p21-waf, and tumor necrosis factor (TNF)-a [36,37], To determine the mechanism of action of proteins, including bcl-2, bax, bcl-X, p21-waf proteins, including bcl-2, bax, bcl-X, p21-waf paclitaxel in ovarian cancer, Ahn et al. [38] tested the effects of the paclitaxel, on the SKOV3 human ovarian carcinoma cell line. While treatment of cells with paclitaxel resulted in bcl-2 phosphorylation and mitochondrial depolarization, cytochrome c was not released and pro-caspase-3 was not activated. Treatment of SKOV3 cells with paclitaxel resulted in the translocation of AIF from the mitochondria to the nucleus via the cytosol. Taken together, these findings suggest that in SKOV3 cells, paclitaxel induces caspase-independent AIF-dependent apoptosis.

In addition to its microtubule stabilizing activity, paclitaxel also stimulates the lipopolysaccharide signaling pathway in murine macrophages, resulting in secretion of the cytokines interleukin 1β and tumor necrosis factor α (TNF-α)

[39-42]. This effect is independent of its ability to stabilize microtubules, because some derivatives of paclitaxel retain microtubule-stabilizing activity but do not stimulate cytokine secretion [43,44]. In human macrophages, paclitaxel alone has not been shown to induce TNF- α or interleukin-1β secretion, but it does enhance production of these cytokines when applied in conjunction with lipopolysaccharide [45].

1.3. Clinical Action

Paclitaxel is one of the most active antineoplastic agents with a wide spectrum of antitumor activity when used as monotherapy [46,47] or in combination chemotherapy regimens. It is used as standard therapy for ovarian, breast, lung, head and neck, and has recognized antitumor activity in several other malignancies.

Paclitaxel is marketed commercially in a formulation that contains a solvent system of the nonionic surfactant Cremophor EL (polyoxyethylated castor oil) and dehydrated ethanol USP (Taxol, Bristol-Myers Squibb Co., Princeton, NY [46]. However, the amount of Cremophor in paclitaxel per administration is relatively high and has been associated with toxic effects by producing or contributing to the well-described hypersensitivity reactions. That commonly occurs during infusion, affecting 25–30% of treated patients [47]. To minimize the incidence and severity of these reactions, premedication with histamine 1 and 2 blockers, as well as glucocorticoids (usually dexamethasone), has become standard practice. The cumulative side effects of dexamethasone used as a premedication may add to treatment-related morbidity and, in some instances, result in early discontinuation of therapy. Cremaphor EL may also contribute to chronic paclitaxel toxic effects, such as peripheral neuropathy. An additional problem arising from the Cremophor and ethanol solvent is the leaching of plasticizers from PVC bags and infusion sets in routine clinical use [47]. Consequently, paclitaxel must be prepared and administered in either glass bottles or non-PVC infusion systems and with in-line filtration.

These problematic issues have spurred interest in the development of taxanes with improved solubility in aqueous solutions. Having in the mind that paclitaxel is water insoluble drug, the attention was paid to the development of nanosuspension dosage forms. In this way, the improved bioavailability of paclitaxel and its nanoparticles can provide targeting capabilities when injected intravenously. Currently available the nanoparticle-based product is

Abraxane® or ABI-007, designed for intravenous administration. This suspension consisting of lyophilized particles (about 130 nm [48]) with 10% (w/w) paclitaxel and 90% (w/w) albumin. To compare the clinical pharmacokinetic properties of paclitaxel formulated as an albumin-bound nanoparticle Abraxane®) and formulated in Cremophorethanol (Taxol) Sparreboom et al. [49] treated 27 patients with advanced solid tumors. The patients were treated with Abraxane® (260 mg/m^2, 3 0 minutes; n = 14) or Taxol (175 mg/m^2, 3 hours; n = 13) with cycles repeated every 3 weeks. The obtained clinical data indicate that paclitaxel plasma clearance and volume of distribution were significantly higher for Abraxane® than for Taxol in humans. Thus, this formulation in compared with Taxol® has rapid infusion rate, absence of requirement for premedication, and a high paclitaxel maximum tolerated dose [50]. The nanoparticle formulation showed a higher response rate, higher time to tumor progression, and lower incidence of grade 4 neutropenia. Additional studies have confirmed that the paclitaxel–albumin nanoparticles can be combined with other chemotherapy agents: carboplatin [51], gemcitabine [52] and squalamine [53].

Also, one of the ways for the increase of the solubility of paclitaxel in aqueous solution is its solubilization in ReGel. ReGel is a non-ionic surfactant that spontaneously forms polymeric micelles containing a hydrophobic core. OncoGel, a controlled-release depot formulation of paclitaxel in ReGel, was designed in order to provide targeted cytotoxicity for local delivery of paclitaxel to solid tumors. Elstad and Fowers [54] during the clinical studies in superficially-palpable tumors and esophageal carcinoma were confirmed local paclitaxel release from OncoGel in patients. OncoGel's ability to improve treatment options of esophageal and brain cancers will be further evaluated.

Fonseca et al. [55] developed a polymeric drug delivery system for paclitaxel, intended to be intravenously administered, aiming at avoiding the use of Cremophor EL and improving the anti-tumoral efficacy of the drug. For this purpose nanoparticles of poly(lactic-co-glycolic acid) containing paclitaxel were prepared by the interfacial deposition method. The *in vitro* anti-tumoral activity of this polymeric drug delivery system was assessed using a human small cell lung cancer cell line (NCI-H69 SCLC) and compared to the commercial formulation Taxol. The obtained results demonstrate that incorporation of paclitaxel in nanoparticles strongly enhances the cytotoxic effect of the drug as compared to Taxol.

Matsumura and Kataoka [56] developed new paclitaxel-incorporating micelle carrier systems (NK105). Micellar nanoparticles are composed of the block copolymers consisting of polyethylene glycol and polyaspartate.

Paclitaxel was incorporated into polymeric micelles formed by physical entrapment utilizing hydrophobic interactions between paclitaxel and the block copolymer polyaspartate chain. During the clinical study, NK105 was administered by intravenous infusion for 1 h every 3 weeks without anti-allergic premedication. Tri-weekly 1h infusion of NK105 was feasible and well tolerated, without allergic reactions. Thus, the incidence of drug-induced side-effects was decreased using paclitaxel in this form due to the reduced drug distribution in normal tissue. The reason for this is because the micelle systems can also evade non-specific capture by the reticuloendothelial system because the outer shell of a micelle is covered with polyethylene glycol. Consequently, a polymer micelle carrier can be delivered selectively to tumor by utilizing the enhanced permeability and retention effect.

It is known that paclitaxel presents a drug of choice in treatment of the endometrial cancer (ovarian and breast cancers). Nowadays, paclitaxel is commonly used in combination with other chemotherapy agents, as well as radiotherapy for the improvement of its efficiency. Paclitaxel exhibits synergistic cytotoxicity with a numberr of cytostatics, including platinum and doxorubicin.

Therefore, the clinical utility of doxorubicin/paclitaxel combination can causes the considerable nonhematological side effects in particular cardiotoxicity and mucotoxicity [57-59]. As better alternative in treatment of breast and ovarian carcinomas, Kurbarcher et al. [60] proposed the mitoxantrone/paclitaxel combination. This combination had a higher clinical activity than doxorubicin/paclitaxel. Thus, this combination showed unusually promising anticancer activity in heavily pretreated patients with platinum-refractory ovarian cancer without producing nontolerated toxicity.

In order to develop a potentially more effective chemotherapy, paclitaxel was combined with a platin analog, carboplatin due to its lesser toxicity [61]. The clinical study was confirmed that the carboplatin-paclitaxel combination is an efficacious, low-toxicity regimen for managing primarily advanced or recurrent endometrial cancers.

Muggia et al. [62] performed Phase III randomized study of cisplatin versus paclitaxel versus cisplatin and paclitaxel in patients with suboptimal stage III or IV ovarian cancer. During the study, the patients receiving cisplatin (100 mg/m^2) or 24-hour infusion paclitaxel (200 mg/m^2) or the paclitaxel/cisplatin combination combination (135/75 mg/m^2). The obtained results indicate that the combination therapy had a better toxicity profile. Therefore, this paclitaxel/cisplatin combination remains the preferred initial treatment option. Unlike this investigation, Ozols et al., [63] conducted a noninferiority trial of cisplatin and paclitaxel versus carboplatin and paclitaxel. The research was conducted on the patients with advanced ovarian cancer after surgery. These patients were randomly assigned to receive cisplatin 75 mg/m^2 plus a 24-hour infusion of paclitaxel 135 mg/m^2 (arm I) or carboplatin area under the curve 7.5 intravenously plus paclitaxel 175 mg/m^2 over 3 hours (arm II). The chemotherapy regimen consisting of carboplatin plus paclitaxel results in less toxicity, is easier to administer, and is not inferior, when compared with cisplatin plus paclitaxel.

In order to evaluate the worth of paclitaxel as adjuvant chemotherapy for patients with resected operable breast cancer and histologically positive nodes, Mamounas et al. [64] designed randomized trial to determine whether the addition of paclitaxel (given at 225 mg/m^2 as a 3-hour infusion) following four cycles of doxorubicin/cyclophosphamide would prolong disease-free survival and overall survival when compared with the administration of four cycles of doxorubicin/cyclophosphamide alone. The results indicate that the addition of paclitaxel to doxorubicin/cyclophosphamide resulted in significant improvement in disease-free survival but no significant improvement in overall survival with acceptable toxicity. No significant interaction between treatment effect and receptor status or tamoxifen administration was observed.

Many published reports have supported the role of paclitaxel in radio sensitization of human tumor cell cultures of different cell types [65-67]. More than 10 different dose schedules of paclitaxel have been tested in clinical trials where the paclitaxel was given concurrently with radiation for the treatment of human solid tumors [68-70]. These clinical trials, which reported different response rates and toxicity profiles, included many Phase I/II clinical trials combining paclitaxel and radiotherapy for lung cancer. The response rates were in the range of 65–86%, but treatment toxicity generally was high, particularly in neutropenia, thrombocytopenia, esophagitis, and pneumonitis [71-73]. This is likely because of the radiosensitizing effect of paclitaxel on normal tissues as well. Because, cell cycle progression is a dynamic process, most clinical trials using paclitaxel concurrently with radiotherapy may not have fully incorporated the best timing for paclitaxel treatment and irradiation to maximize the G2-M cell cycle effect for radiosensitization. The optimal dose schedule in balancing the radiosensitization effect with the efficacy of therapy for paclitaxel and concurrent radiation treatment for lung cancer has not been defined.

Chen et al. [74] present in their article a Phase I/II clinical study using pulsed low dose paclitaxel and radiation for thoracic malignancy. The study was based on preclinical data. Having in the mind, Chen et al. hypothesized that maximal tumor control could be achieved using pulsed low-dose paclitaxel with radiation if the timing of radiation was synchronized to allow for the maximal cell cycle and apoptotic effects of paclitaxel. During the Phase I/II clinical study for non small cell lung cancer used pulsed low-dose paclitaxel three times/week with the starting dose of 15 mg/m^2. Daily thoracic radiotherapy was delivered in 1.8 Gy/fraction to 60–65 Gy for gross disease and to 45–58 Gy for microscopic disease. Timing of radiotherapy was delayed to allow for a minimum of 4 h for cell cycle progression. On the basis of the clinical researches, it can be said that the pulsed low-dose paclitaxel radiosensitization for non-small cell lung cancer resulted in a superior local control rate and comparable survival rate when compared with chemoradiation regimens using systemic dose chemotherapy. The regimen is associated with low toxicity.

In addition to the treatment of cancer, paclitaxel was taken the important place in the treatment of long, complex coronary artery lesions. Using the drug-eluting stents providing local drug delivery at the time of stent implantation reduced indices of restenosis, including the need for reintervention, compared with bare metal stents. The system of paclitaxel- and sirolimus-eluting stents has become widely used in clinical practice. Dawkins et al. [75] estimated that the polymer-based paclitaxel-eluting (TAXUS Moderate Release) stent system is safe and effective in treatment of long, complex coronary artery lesions and also provides the evidence base for the more widespread use of drug-eluting stents in contemporary clinical practice.

Recent studies showed that paclitaxel is able to induce early reactive oxygen species (ROS) production in cancer cells. Also, hydrogen peroxide (H_2O_2) was found to be involved in paclitaxel-induced cancer cell death *in vitro* and *in vivo* [76-78]. The paclitaxel-induced ROS generation was occurred rapidly within several hours of drug exposure, with $O_2^-$and H_2O_2 accumulation mainly outside the cells while the intracellular ROS remained unchanged. The increase in extracellular ROS caused lethal damage to the bystander cancer cells not exposed to paclitaxel. The studies are focused in the direction of the modification of paclitaxel structure in order to improve its clinical applications due the side effects. Also, the more reason for its modification is to improve the solubility characteristics. Therefore, the complexity of the paclitaxel structure presents a major obstacle to facile chemical modification. In recent years, considerable academic and industrial research effort had been focused on the discovery of agents that act mechanistically like the paclitaxel, but with improved biological or pharmaceutical characteristics. Three other new classes of natural products have recently been identified that share the same antimitotic mechanism of action as the taxanes, *i.e.,* eleutherobin and the structurally related sarcodictyins, discodermolide, and the epothilones.

Conclusion

Paclitaxel presents the most well-known natural-source cancer drug that was proved as very efficient for the treatmnent of various cancers especially ovarian and breast cancer. The synthetic pathway of obtaining paclitaxel is very complex and no efficient, so that the isolation from the natural source (*Taxus brevifolia*) is the only way for production of this substance. Having in the mind this fact, the knowledge of the paclitaxel biosynthesis consisting from 20 enzymatic steps is necessary.

Based on the studies of its mechanism of action, it can be concluded that this drug alters microtubule assembly inhibiting microtubule depolymerization and changing microtubule dynamics. These effects cause the disruption of normal reorganization of the microtubule network, which is necessary for mitosis and cell proliferation. Paclitaxel has limited application in the intravenous administration due to its poor water solubility. From these reasons, the attention of researchers is focused in the development of alternate formulation of paclitaxel having good aqueous solubility and at the same time free of any side effects. The various approaches employed so far include cosolvents, emulsions, micelles, liposomes, microspheres nanoparticles, cyclodextrins, pastes and implants.

Acknowledgments

This work was supported by the Ministry of Education, Science and Technological Development of the Republic of Serbia under the project TR-34012. Ivan Savic is a recipient of a scholarship granted by the Ministry of Education, Science and Technological Development of the Republic of Serbia.

References

[1] Nirmala, M. J., Samundeeswari, A. & Deepa Sankar, P. (2011). Natural plant resources in anti-cancer therapy-A review. *Res. Plant Biol.*, *1(3)*, 1-14.

[2] Rixe, O. & Fojo, T. (2007). Is cell death a critical end point for anticancer therapies or is cytostasis sufficient?. *Clin. Cancer Res.*, *13(24)*, 7280-7287.

[3] Hartwell, J.L. *Plants Against Cancer: A Survey*. Lawrence; Quarterman Publications; 1982.

[4] Nikolic, V. D., Savic, I. M., Savic, I. M., Nikolic, Lj., Stankovic, M. & Marinkovic, V. (2011). Paclitaxel as an anticancer agent: isolation, activity, synthesis and stability. *Cent. Eur. J. Med.*, *6(5)*, 527-536.

[5] Schiff, P. B., Fant, J. & Horwitz, S. B. (1979). Promotion of Microtubule Assembly In Vitro by Taxol. *Nature*, *277*, 655-667.

[6] Jennewein, S. & Croteau, R. (2001). Taxol: biosynthesis, molecular genetics, and biotechnological applications. *Appl. Microbiol. Biotechnol.*, *57*, 13–19.

[7] Singla, A. K., Garg, A. & Aggarwal, D. (2002). Paclitaxel and its formulations *Int. J. Pharmaceut.*, *235*, 179–192.

[8] Hezari, M. & Croteau, R. Taxol Biosynthesis: An Update. (1997). *Planta Medica*, *63*, 291-295.

[9] Hartzell Jr. H. *The Yew Tree*. Eugene, Oregon; Hulogosi Press; 1991.

[10] Horwitz, S. B. (1992). Mechanism of action of taxol. *Trends Pharmacol. Sci.*, *13*, 134-136.

[11] Jordon, M. A. & Wilson, L. In: *Taxane anticancer agents: Basic science and current status*. Georg, C. I., Chen, T., Ojima, I. & Vyas, D. M. editors. Washington: American Chemical Society; 1995; 138-153.

[12] Borman, S. (1994). Total synthesis of anticancer agent taxol achieved by 2 different routes. *Chem. Eng. News*, *72*, 32–34.

[13] Koepp, A. E., Hezari, M., Zajicek, J., Stofer Vogel, B., LaFever R. E., Lewis, N. G. & Croteau, R. (1995). Cyclization of geranylgeranyl diphosphate to taxa-4(5),11(12)-diene is the committed step of Taxol biosynthesis in Pacific yew. *J. Biol. Chem.*, *270*, 8686–8690.

[14] Hezari, M., Lewis, N. G. & Croteau, R. (1995). Purification and characterization of taxa-4(5),11(12)-diene synthase from Pacific yew (Taxus brevifolia) that catalyses the first committed step of Taxol biosynthesis. *Arch. Biochem. Biophys.*, *322*, 437–444.

[15] Hezari, M., Ketchum, R. E. B., Gibson, D. M. & Croteau, R. (1997). Taxol production and taxadiene synthase activity in Taxus Canadensis cell suspension cultures. *Arch. Biochem. Biophys.*, *337*, 185–190.

[16] Lin, X., Hezari, M., Koepp, A. E., Floss, H. G. & Croteau, R. (1996). Mechanism of Taxadiene Synthase, a Diterpene Cyclase That Catalyzes the First Step of Taxol Biosynthesis in Pacific Yew. *Biochemistry*, *35*, 2968-2977.

[17] Hefner, J., Rubenstein, S. M., Ketchum, R. E. B., Gibson, D. M., Williams, R. M. & Croteau, R. (1996). Cytochrome P450-catalyzed hydroxylation of taxa-4(5),11(12)-diene to taxa-4(20),11(12)-dien-5α-ol: The first oxygenation step in Taxol biosynthesis. *Chem. Biol.*, *3*, 479–489.

[18] Croteau, R., Hezari, M., Hefner,J., Koepp, A. & Lewis, N. G. In: Taxane anticancer agents: Basic science and current status. Georg, G. I., Chen, T., Ojima, I. & Vyas, D. M. editors. Washington; American Chemical Society; 1995; 72—80.

[19] Hefner, J. & Croteau, R. (1996). Microsome preparation from woody plant tissues. *Methods Enzymol.*, *272*, 243-250.

[20] Mihaliak, C. A., Karp, F. & Croteau, R. (1993). Cytochrome P450 terpene hydroxylases. *Methods Plant Biochem.*, *9*, 261—279.

[21] Kingston, D. G. I., Molinero, A A. & Rimoldi, J. M. (1993). The taxane diterpenoids. *Prog. Chem. Org. Nat. Prod.*, *61*, 1-206.

[22] Gueritte-Voegelein, F., Guenard, D. & Potier, P. (1987). Taxol and derivatives: a biogenetic hypothesis. *J. Nat. Prod.*, *50*, 9-18.

[23] Floss, H. C. & Mocek, U. In: *Taxol: Science and applications*. Suffness, M. editor. Boca Raton; CRC Press; 1995; 191-208.

[24] *Croteau, R., Hefner, J., Hezari, M. & Lewis, N. G.* (1995). Taxol biosynthesis: cyclization and early hydroxylation steps of the pathway. *Curr. Top Plant Physiol.*, 15, 94–104.

[25] Desai, A. & Mitchison, T. J. (1997). Microtubule polymerization dynamics. *Annu. Rev. Cell Dev. Biol.*, *13*, 83-117.

[26] Lee, F. Y. F., Borzilleri, R., Fairchild, C. R., Kim, S. H., Long, B. H., Reventos-Suarez, C., Vite, G. D., Rose, W. C. & Kramer, R. A. (2001). BMS-247550: A Novel Epothilone Analog with a Mode of Action Similar to Paclitaxel but Possessing Superior Antitumor Efficacy. *Clin. Cancer Res.*, *7*, 1429–1437.

[27] Schiff, P. B. & Horwit, S. B. (1980). Taxol stabilizes microtubules in mouse fibroblast cells. *Proc. Nati. Acad. Sci. USA*, *77*, 1561-1565.

[28] Parness, J. & Horwitz, S. B. (1981). Taxol binds to polymerized tubulin in vitro. *J. Cell Biol.*, *91*, 479-487.

[29] Schiff, P. B. & Horwitz, S. B. (1981). Taxol assembles tubulin in the absence of exogenous guanosine 5'-triphosphate or microtubule-associated proteins. *Biochemistry*, *20*, 3247-3252.

[30] Hamel, E., del Campo, A. A., Lowe, M. C. & Lin, C. M. (1981). Interactions of taxol, microtubule-associated proteins, and guanine nucleotidcs in tubulin polymerization. *J. Biol. Chem.*, *256*, 11887-11894.

[31] DeBrabander, M., Geuens, G., Nuydens, R., Willebrords, R. & DeMey, J. (1981). Taxol induces the assembly of free microtubules in living cells and blocks the organizing capacity of the centrosomes and kinetochores. *Proc. Nati. Acad. Sci. USA*, *78*, 5608-5612.

[32] Long, B. H., Carboni, J. M., Wasserman, A. J., Cornell, L. A., Maria Casazza, A., Jensen, P. R., Lindel, T., Fenica, W. & Fairchild, C. R. (1998). Eleutherobin, a Novel Cytotoxic Agent That Induces Tubulin Polymerization, Is Similar to Paclitaxel (Taxola®), *Cancer Res.*, *58*, 1111-1115.

[33] Schrijvers, D., Wanders, J., Dirix, L., Prove, A., Vonck Van Ooslerom. A., & Kaye, S. (1993). Coping with toxicities of docetaxel (Taxotere'M). *Ann Oncol*, *4*, 610-611.

[34] Jordan, M. A., Toso, R. J., Thrower, D. & Wilson, L. (1993). Mechanism of milotic block and inhibition of cell proliferation by taxol at low concentrations. *Proc. Nati. Acad. Sci. USA*, *90*, 9552-9556.

[35] Fan, W. (1999). Possible mechanisms of paclitaxel-induced apoptosis. *Biochem. Pharmacol.*, *57*, 1215–1221.

[36] Haldar, S., Chintapalli, J. & Croce, CM. (1996). Taxol induces bcl-2 phosphorylation and death of prostate cancer cells. *Cancer Res.*, *56*, 1253–1255.

[37] Tudor, G., Aguilera, A., Halverson, D. O., Laing, N. D. & Sausville, E. A. (2000). Susceptibility to drug-induced apoptosis correlates with differential modulation of Bad, Bcl-2, and Bcl-xL protein levels. *Cell Death Differ.*, *7*, 574–586.

[38] Ahn, H. J., Kim, Y., Kim, J. U., Han, S. M., Shin, J. W. & Yang, H. O. (2004). Mechanism of Taxol-Induced Apoptosis in Human SKOV3 Ovarian Carcinoma Cells. *J. Cell Biochem.*, *91*, 1043–1052

[39] Ding, A. H., Porteau, F., Sanchez, E. & Nathan, C. F. (1990). Shared actions of endotoxin and taxol on TNF receptors and TNF release. *Science*, *248*, 370–372.

[40] Bogdan, C. & Ding, A. (1992). Taxol, a microtubule-stabilizing antineoplastic agent, induces expression of tumor necrosis factor alpha and interleukin-1 in macrophages. *J. Leukocyte Biol*, *52*, 119–121.

[41] Manthey, C., Brandes, M., Perera, P. & Vogel, S. (1992). Taxol increases steady-state levels of lipopolysaccharide-inducible genes and protein-tyrosine phosphorylation in murine macrophages. *J. Immunol*, *149*, 2459–2465.

[42] Carboni, J., Singh, C. & Tepper, M. (1993). Taxol and lipopolysaccharide activation of a murine macrophage cell line and induction of similar tyrosine phosphoproteins. *Natl. Cancer Inst. Monogr*, *15*, 95–101.

[43] Manthey, C., Qureshi, N., Stutz, P. & Vogel, S. (1993). Lipopolysaccharide antagonists block taxol-induced signaling in murine macrophages. *J. Exp. Med.*, *178*, 695–702.

[44] Burkhart, C., Berman, J., Swindell, C. & Horwitz, S. (1994). Relationship between the structure of taxol and other taxanes on induction of tumor necrosis factor-alpha gene expression and cytotoxicity. *Cancer Res.*, *54*, 5779–5782.

[45] Allen, J., Moore, S. A. & Wewers, M. D. (1993). Taxol enhances but does not induce interleukin-1 beta and tumor necrosis factor-alpha production. *J. Lab. Clin. Med.*, *122*, 374–381.

[46] Woo, H. L., Swenerton, K. D. & Hoskins, P. J. (1996). Taxol is active in platinum resistant endometrial adenocarcinoma. *Am. J. Clin. Oncol.*, *19*, 290-291.

[47] Ball, H. G., Blessing, J. A., Lentz, S. S. & Mutch, D. G. (1996). A phase II trial of paclitaxel in patients with advanced or recurrent adenocarcinoma of the endometrium: A Gynecologic Oncology Group study. *Gynecol. Oncol.*, *62*, 278-281.

[48] Nyman, D., Campbell, K. Hersh, E. Long, K., Richardson, K., Trieu, V., Desai, N. Hawkins, M. & von Hoff, D. (2005). Phase I and harmacokinetics trial of ABI-007, a novel nanoparticle formulation of paclitaxel in patients with advanced nonhematologic malignancies. *J. Clin. Oncol.*, *1*, 7785–7793.

[49] Sparreboom, A., Scripture, C. D., Trieu, V., Williams, P. J., De, T., Yang, A., Beals, B., Figg, W. D., Hawkins, M. & Desai, N. (2005). Comparative Preclinical and Clinical Pharmacokinetics of a Cremophor-Free, NanoparticleAlbumin-Bound Paclitaxel (ABI-007) and Paclitaxel Formulated in Cremophor (Taxol). *Clin. Cancer Res.*, *11*, 4136-4143.

[50] Ibrahim, N., Desai, N., Legha, S., Soon-Shiong, P., Theriault, R., Rivera, E., Esmaeli, B., Ring, S., Bedikian, A., Hortobagyi, G. & Ellerhorst, J. (2002). Phase I and pharmacokinetic study of ABI-007, a Cremophor-free, proteinstabilized, nanoparticle formulation of paclitaxel. *Clin. Cancer Res.*, *8*, 1038–1044.

[51] Stinchcombe, T., Socinski, M., Walko, C., O'neil, B., Collichio, F., Ivanova, A., Mu, H., Hawkins, M., Goldberg, R., Lindley, C. & Claire Dees, E. (2007). Phase I and pharmacokinetic trial of carboplatin and albumin-bound paclitaxel, ABI-007 (Abraxane®) on three treatment schedules in patients with solid tumors. *Cancer Chemoth Pharmacol.*, *60*, 759–766.

[52] Moreno-Aspitia, A. & Perez, E. (2005). North Central Cancer Treatment Group N0531: phase II trial of weekly albumin-bound paclitaxel (ABI-007; Abraxane®) in combination with gemcitabine in patients with metastatic breast cancer. *Clin. Breast Cancer*, *6*, 361–364.

[53] Herbst, R. S., Hanmmond, L. A., Carbone, K. J., Tran, H. T., Holroyd, K. J., Desai, A., Williams, J. I., Bekele, B. N., Hait, H., Allgood, V., Solomon, S. & Schiller, J. H. (2003). A Phase I/II Trial of continuous five-day infusion of squalamine lactate (MSI-1256F) plus carboplatin and paclitaxel in patients with advanced non-small lung cancer. *Clin. Cancer Res.*, *9*, 4108-4115.

[54] Elstad, N. L. & Fowers, K. D. (2009). OncoGel (ReGel/paclitaxel) - Clinical applications for a novel paclitaxel delivery system. *Adv. Drug Deliver Rev.*, *61*, 785–794.

[55] Fonseca, C., Simoes, S. & Gaspar, R. (2002). Paclitaxel-loaded PLGA nanoparticles: preparation, physicochemical characterization and in vitro anti-tumoral activity. *J. Control Release*, *83*, 273–286.

[56] Matsumura, Y. & Kataoka, K. (2009). Preclinical and clinical studies of anticancer agent-incorporating polymer micelles. *Cancer Sci.*, *100(4)*, 572–579.

[57] McGuire, W. P., Rowinsky, E. K., Rosenshein, N. B., Grumbine, F. C., Ettinger, D. S., Armstrong, D. K. & Donehower, R.C. (1989). Taxol: a unique antineoplastic agent with significant activity in advanced ovarian neoplasms. *Ann. Intern Med.*, *111*, 273-279.

[58] Riggs, C. E. Antitumor antibiotics and related compounds. In: Perry, M. C. editor. *The Chemoterapy Source Book.*, Baltimore: Williams & Wilins; 1992; 318-358.

[59] Onetto, N., Dougan, M., Hellman, S., Gustafson, N., Borroughs, J., Florczyk, A., Canetta, R. & Rozenczweig, M. Safety profile. In:

McGuire W. P. & Rowinsky, E. K. editors. *Paclitaxel in Cancer Treatment*. New York: Marcel Dekker, Inc.; 1995; 121-149.

[60] Kurbacher, C. M., Bruckner, H. W., Cree, I. A., Kurbacher, J., Wilhelm, L., Poch, G., Indefrei, D., Mallmann, P. & Andreotti, P. (1997). Mitoxantrone combined with paclitaxel as salvage therapy for platinum-refractory ovarian cancer: laboratory study and clinical pilot trial. *Clin. Cancer Res.*, *3*, 1527-1533.

[61] Hoskins, P. J., Swenerton, K. D., Pike, J. A., Wong, F., Lim, P., Acquino-Parsons, C. & Lee, N. (2001). Paclitaxel and Carboplatin, Alone or With Irradiation, in Advanced or Recurrent Endometrial Cancer: A Phase II Study. *J. Clin. Oncol.*, *19(20)*, 4048-4053.

[62] Muggia, F. M., Braly, P. S., Brady, M. F., Sutton, G., Niemann, T. H., Lentz, S. L., Alvarez, R. D., Kucera, P. R. & Small, J. M. (2000). Phase III Randomized Study of Cisplatin Versus Paclitaxel Versus Cisplatin and Paclitaxel in Patients With Suboptimal Stage III or IV Ovarian Cancer: A Gynecologic Oncology Group Study. *J. Clin. Oncol.*, *18(1)*, 106-115.

[63] Ozols, R. F., Bundy, B. N., Greer, B. E., Fowler, J. M., Clarke-Pearson, D., Burger, R. A., Mannel, R. S., DeGeest, K., Hartenbach, E. M. & Baergen, R. (2003). Phase III Trial of Carboplatin and Paclitaxel Compared With Cisplatin and Paclitaxel in Patients With Optimally Resected Stage III Ovarian Cancer: A Gynecologic Oncology Group Study. *J. Clin. Oncol., 21(17)*, 3194-3200.

[64] Mamounas, E. P., Bryant, J., Lembersky, B., Fehrenbacher, L., Sedlacek, S. M., Fisher, B., Wickerham, D. L., Yothers, G., Soran, A. & Wolmark, N. (2005). Paclitaxel After Doxorubicin Plus Cyclophosphamide As Adjuvant Chemotherapy for Node-Positive Breast Cancer: Results From NSABP B-28, *J. Clin. Oncol.*, *23*, 3686-3696.

[65] Rodriguez, M., Sevin, B. U., Perras, J., Nguyen, H. N., Pham, C., Steren, A. J., Koechli, O. R. & Averette, H. E. (1995). Paclitaxel: A radiation sensitizer of human cervical cancer cells. *Gyn. Oncol.*, *57*, 165–169.

[66] Lokeshwar, B. L., Ferrell, S. M. & Block, N. L. (1995). Enhancement of radiation response of prostatic carcinoma by Taxol: therapeutic potential for late-stage malignancy. *Anticancer Res.*, *15*, 93–98.

[67] Elomaa, L., Joensuu, H., Kulmala, J., Klemi, P. & Grenman, R. (1995). Squamous cell carcinoma is highly sensitive to Taxol, a possible new radiation sensitizer. *Acta Otolaryngol.*, *115:* 340–344.

[68] Lau, D., Ryu, J., Gandara, D., Morgan, R., Doroshow, J., Wilder, R. & Leigh, B. (1999). Concurrent twice-weekly paclitaxel and thoracic irradiation for stage III non-small cell lung cancer. *Semin. Radiat. Oncol.*, *9*, 117–120.

[69] Kirkbride, P., Gelmon, K., & Eisenhauer, E. (1999). Paclitaxel and concurrent radiotherapy in locally advanced non-small cell lung cancer: the Canadian experience. *Semin. Radiat. Oncol.*, *9*, 102–107.

[70] Formenti, S. C., Symmans, W. F., Volm, M., Skinner, K., Cohen, D., Spicer, D. & Danenberg, P. V. (1999). Concurrent paclitaxel and radiation therapy for breast cancer. *Semin. Radiat. Oncol.*, *9*, 34–42.

[71] Choy, H., Yee, L. & Cole, B. F. (1995). Combined-modality therapy for advanced non-small cell lung cancer: Paclitaxel and thoracic irradiation. *Semin. Oncol.*, *22*, 38–44.

[72] Antonia, S. J., Wagner, H., Williams, C., Alberts, M., Hubbell, D., Robinson, L., Hilstro, J. & Ruckdeschel, J. C. (1995). Concurrent paclitaxel/cisplatin with thoracic radiation in patients with stage IIIA/B non-small cell carcinoma of the lung. *Semin. Oncol.*, *22*, 34–37.

[73] Greco, F. A., Stroup, S. L., Gray, J. R. & Hainsworth, J. D. (1996). Paclitaxel in combination chemotherapy with radiotherapy in patients with unresectable stage III non-small-cell lung cancer. *J. Clin. Oncol.*, *14*, 1642–1648.

[74] Chen, Y., Pandya, K., Keng, P. C., Johnstone, D., Li, J., Lee, Y. J., Smudzin, T. & Okunieff, P. (2003). Phase I/II Clinical Study of Pulsed Paclitaxel Radiosensitization for Thoracic Malignancy: A Therapeutic Approach on the Basis of Preclinical Research of Human Cancer Cell Lines. *Clin. Cancer Res.*, *9*, 969-975.

[75] Dawkins, K. D., Grube, E., Guagliumi, G., Banning, A. P., Zmudka, K., Colombo, A., Thuesen, L., Hauptman, K., Marco, J., Wijns, W., Popma, J. J., Koglin, J. & Russell, M. E. (2005). Clinical Efficacy of Polymer-Based Paclitaxel-Eluting Stents in the Treatment of Complex, Long Coronary Artery Lesions From a Multicenter, Randomized Trial Support for the Use of Drug-Eluting Stents in Contemporary Clinical Practice. *Circulation*, *112*, 3306-3313.

[76] Parness, J., & Horwitz, S. B. (1981). Taxol binds to polymerized tubulin *in vitro*. *J. Cell Biol.*, *91*, 479–487.

[77] Manfredi, J. J., Parness, J., & Horwitz, S. B. (1982). Taxol binds to cellular microtubules. *J. Cell Biol.*, *94*, 688–696.

[78] Kumar, N. (1981). Taxol-induced polymerization of purified tubulin. Mechanism of action. *J. Biol. Chem.*, *256*, 10435–10441.

In: Paclitaxel
Editor: Diego Morales
ISBN: 978-1-62808-549-5

Chapter 4

Current Status and Clinical Trials of Paclitaxel-Containing Therapy for Urothelial Cancer

Yasuyoshi Miyata and Hideki Sakai*
Department of Nephro-Urology, Nagasaki University Graduate School of Biomedical Sciences, Nagasaki, Japan

Abstract

Urothelial cancer (UC) arising from the urinary bladder, renal pelvis, or ureter is a common cancer in adult males. In general, UC is recognized as chemo-sensitive and patients with advanced-stage lesions including metastatic UC are treated using chemotherapy-based methods. At present, combination therapy with gemcitabine and cisplatin (CDDP) is standard in patients with such advanced UC. MVAC therapy comprising methotrexate, vinblastine, doxorubicin, and CDDP is also used for these patients. However, unfortunately, most of these patients experience relapse and no standard salvage therapies have been devised for use after CDDP-containing regimens. In recent years, various new chemotherapeutic agents and regimens have been reported as effective

* Corresponding author: Yasuyoshi Miyata, MD, PhD, Department of Nephro-Urology, Nagasaki University Graduate School of Biomedical Sciences, 1-7-1 Sakamoto, Nagasaki 852-8501, Japan., Phone: +81 95 849 7340, Fax: +81 95 849 7343, E-mail: int.doc.miya@m3.dion.ne.jp.

and safe for CDDP-resistant UC, and various targeted and anti-angiogenic therapies are also performed. However, most of these approaches have been described in preliminary studies and further investigations are necessary to confirm their clinical feasibility. Clinical studies are thus in progress all over the world. Paclitaxel (PTX) appears effective against UC. Previous reports have identified PTX monotherapy as an active first line chemotherapeutic regimen. In addition, this agent is considered a good option for salvage therapy, because it is well tolerated and safe. On the other hand, whether PTX monotherapy or PTX-based combination chemotherapy offers satisfactory anti-tumoral effects and prolonged survival in patients with CDDP-resistant UC remains unclear. The aim of this study was to clarify the current status of PTX-containing chemotherapy as second line therapy for patients with UC. In addition, we introduce clinical trials of PTX-containing chemotherapies and combination therapies with PTX and molecular-targeted therapy for advanced UC. We also describe delivery systems and intravesical therapies using PTX. This chapter therefore introduces comprehensive and wide-ranging information regarding the current status of PTX-based therapies in patients with UC.

Introduction

Urothelial cancer (UC) of the urinary bladder is the fourth-most common cancer in men and approximately a quarter of patients show advanced disease at diagnosis [1]. While UC of the upper urinary tract is relatively rare, it has a high frequency of metastasis due to the upper urinary tract's anatomical characteristics like a thin muscle layer, proximity to the kidneys, and rich lymphatic drainage. In addition, UC is recognized to have high malignant potential, and the median survival of patients with metastatic disease who receive no aggressive treatment is 4-6 months. Discussion of treatment strategies for patients with advanced UC is thus essential.

In almost all cancers, chemotherapy represents a common and important approach to treatment, particularly in advanced-stage and metastatic disease. In general, UC arising from the upper urinary tract or urinary bladder is recognized as chemo-sensitive [2]. Among various chemotherapeutic agents, cisplatin (CDDP) is considered the most useful and effective against UC, as with various other malignancies. In the 1990s, the "gold standard" chemotherapy for the treatment of patients with advanced UC was the MVAC regimen comprising methotrexate, vinblastine, doxorubicin, and CDDP, among various CDDP-containing regimens [3,4]. This regimen is effective compared to cisplatin monotherapy in patients with metastatic UC, but is relatively toxic and shows a high frequency of severe adverse events [5] . That is, response rates for MVAC as a first line therapy for treating advanced UC are reportedly approximately 70%, but the incidence of drug-related high-grade toxicity including death is approximately 60% [3-5]. Preventing and reducing adverse events is thus problematic with the MVAC regimen.

In 2000, the combination of gemcitabine (GEM) and CDDP was reported in a randomized phase III study to show similar anti-tumoral efficacy to the MVAC regimen, but with lower toxicity [6]. Actual frequencies of neutropenic fever with this CG regimen and the MVAC regimen were 2% and 14%, respectively. In addition, the mortality rate due to toxicity was 1% on the GC regimen and 3% on the MVAC regimen. The GC regimen is thus currently considered the gold standard in patients with advanced UC. These CDDP-based combination chemotherapies are associated with clinical remission in about half of advanced UC patients [7] , and such regimens are commonly used as first line chemotherapy.

Unfortunately, most patients responding to CDDP-based chemotherapies develop disease progression, with the cancer often becoming highly aggressive. In fact, almost all patients experience disease recurrence within the first year, and median survival is approximately 1 year [6,8]. As a result, many urologists, medical oncologists, and researchers pay special attention to second line chemotherapies, but approved and established therapeutic options remain lacking.

In addition to such failings, CDDP-based regimens are unavailable for treatment of some patients because of their nephrotoxicity and other comorbidities. Carboplatin is a platinum-based organic drug that has been used as a substitute for CDDP in various combinations because of its lower nephrotoxicity [2,9]. However, the anti-tumoral effects of carboplatin are generally recognized as inferior to those of CDDP in UC patients. Similar results have also been reported between CDDP- and carboplatin-based chemotherapies for the treatment of advanced UC [10]. That is, overall response and complete response rates with CDDP-based regimens were 36-71% and 13-25%, respectively, whereas rates with carboplatin-based regimens were 28-56% and 0-11%, respectively. Aside from these relative merits between regimens, the anti-tumoral efficacy of carboplatin-based regimens remains unsatisfactory.

Development and clinical trials of non-cisplatin-based chemotherapeutic agents and non-CDDP-based regimens are thus important to improve the outcomes of patients with advanced UC. In particular, new treatment strategies and regimens are becoming increasingly important for improving quality of life (QOL) and prolonging survival in patients with CDDP-resistant UC. This chapter pays special attention to paclitaxel (PTX)-based regimens for UC patients, especially in terms of drug delivery system, combination with molecular-targeted agents and intravesical PTX therapy. In addition, our data on changes in QOL score using PTX-based therapy are also introduced. On the other hand, we highlight the lack of basic information on the therapeutic efficacy and safety of conventional PTX-based regimens needed to make routine decisions regarding treatment.

Why PTX?

Several anti-tumoral agents have been reported as useful and safe for the treatment of UC patients. However, unfortunately, such monotherapies have been barely satisfactory in terms of response rate and patient outcomes, particularly for advanced UC. For example, response rates to CDDP monotherapy and carboplatin monotherapy in advanced UC patients are 12-17% and 12-14%, respectively [11,12]. GEM has been extensively studied using various schedules and different settings, and response rates for GEM monotherapy in patients with advanced UC have been reported as approximately 27-29% [13]. On the other hand, PTX monotherapy showed a relatively high response rate (42%) without severe adverse events for patients with advanced UC [14]. PTX thus has the strongest anti-tumoral effect among investigated anti-cancer agents.

PTX exhibits cytotoxic activities via cell cycle arrest and the regulation of apoptotic pathways [15,16], exerting similar anti-tumoral functions to other types of chemotherapeutic agent. On the other hand, PTX can also function as a microtubule inhibitor that acts by promoting tubulin polymerization and stabilization of microtubules [17], representing a unique characteristic. Microtubules play essential roles in cell proliferation and mitosis in cancer cells [18,19]. As a result, anti-microtubule functions lead to G2-M arrest and mitotic cell death [18]. These unique mechanisms represent desirable characteristics for combination with other types of anti-tumoral agent.

PTX-Based Regimens

As mentioned above, PTX monotherapy offers a relatively high response rate as first line chemotherapy, but the response rate for previously treated advanced UC was just 7% [20]. Actually, to the best of our knowledge, clinical studies and case reports have not defined PTX monotherapy as useful or effective as a second line chemotherapy for UC patients. There is thus general agreement that PTX should be used in combination chemotherapy for advanced and/or pre-treated UC.

Representative PTX-based combination therapies include PTX with CDDP and PTX with carboplatin. However, typical response rates are reportedly 35-65% (median survival, 10-12 months), similar to MVAC and the GC regimen [14, 21]. Another representative PTX-based therapy is the GP regimen, combining PTX with GEM. This GP regimen has interesting pharmacokinetic characteristics, in that PTX increases the accumulation of GEM triphosphate as the active metabolite of GEM [22]. In fact, several studies on the GP regimen as a second line chemotherapy have reported favorable results [23,24]. The M-TEC regimen, comprising methotrexate, PTX, epirubicin, and carboplatin, is similar to the MVAC regimen. This regimen has been applied as first line treatment for urinary bladder metastases [25]. In recent years, the anti-tumor effects and toxicity of this regimen as second line treatment after GC therapy were reported by Halim et al.[26]. This prospective phase II study was performed using methotrexate (40 mg/m^2 on day 15), PTX (175 mg/m^2 on day 1), epirubicin (40 mg/m^2 on day 15), and carboplatin (area under the curve, 5) in 40 patients. The overall response rate was 39% and median progression-free and overall survival rates were 12% and 35%, respectively. Finally, the authors concluded that this regimen offers a modest response rate with acceptable toxicity.

In addition to PTX-based combination therapies, sequential therapy with GEM and carboplatin followed by PTX has been reported as a safer first line treatment for 27 patients with advanced UC [27]. This regimen showed a response rate of 40.7% and 3 patients displayed grade 3 neutropenia and anemia as main toxicities. They concluded that this sequential approach seems safe and useful, but failed to improve outcomes due to the small number of patients. Such sequential therapy is expected to improve the anti-tumoral effects and prolong survival for previously treated patients with advanced UC.

Nano-Particle and Drug Delivery System for PTX

PTX is a hydrophobic molecule with poor solubility in water. The micelle-forming vehicle Cremophor EL, a polyethoxylated castor oil, has therefore been used as a solvent. In current commercial formulation, PTX is dissolved in a 1:1 (v/v) mixture of Cremophor EL and dehydrated alcohol. However, this mixture is toxic and can induce side effects such as hypersensitivity reactions and neurotoxicity [28]. In fact, about a third of patients displayed minor hypersensitivity reactions despite routine pre-medication [29]. This infusion therefore needs a prolonged administration time of 3 h and premedication with corticosteroids and antihistamines. In addition, PTX must be prepared in a non-polyvinyl chloride infusion set with an in-line filter to prevent leaching of di-(2-ethylhexyl) phthalate (DHPE) from Cremophor EL, which is a potential hepatotoxin and mutagen [30,31]. To address such problems and unfavorable issues, various studies and trials have been undertaken. In this section, we show two representative strategies, namely nab-PTX and a drug delivery system.

Overview of Nab-PTX

Nab-PTX is a novel albumin-bound nanoparticulate form of PTX. With this modification, nab-PTX has characteristics of increased solubility and drug delivery to cancer cells via biological interactions with albumin receptors to facilitate drug transport across epithelial cells [32,33]. Several studies have shown that nab-PTX offers highly potent inhibition of cell proliferation in various cancers, both in vitro and in vivo [34]. This formulation was approved by the Food and Drug Administration (FDA) in 2005 as a systematic therapy for treating breast cancers that fail to respond to combination chemotherapy for metastatic disease or show relapse within 6 months of adjuvant chemotherapy. In addition, nab-PTX has advantages in terms of a reduced risk of hypersensitivity and eradication of the need for special intravenous tubing. Finally, nab-PTX does not require premedication or prolonged infusion times.

Arterial infusion of chemotherapy has been used for the treatment of local advanced bladder cancer. Interestingly, one phase I study found that hepatic arterial infusion of nab-PTX is safe at doses up to 260 mg/m^2 [35]. In addition, hepatic arterial infusion of nab-PTX in combination with intravenous GEM and bevacizumab was well tolerated and had anti-tumoral effects for patients with liver metastases [36]. As mentioned above, nab-PTX showed a reduced risk of various adverse events, including hypersensitivity reactions. Various methods of administration are thus possible.

Delivery System

Intra-tumoral delivery of agents is a consistent strategy to provide drug localization within the tumor and minimize exposure of non-target organs. As drug carriers, various types of natural and synthetic polymers have been proposed. However, among these, few have progressed to in vivo or clinical trials, including polymer-based carriers such as in situ gel-forming polymers, thermosensitive polymers, and poly (organophosphazene) hydrogels. What has been investigated is thermosensitive poly (organophosphazene) gel chemically conjugated with PTX [37]. In addition, a PTX-poly (organophosphazene) hydrogel mixture that can be simply prepared by physically mixing PTX with polymer solution showed that this system can accommodate a large dose of the drug in human tumor xenografts [38].

Other investigators have reported that nanoparticle-mediated co-delivery of PTX and a toll-like receptor 4 agonist may suppress tumor growth and enhance immune response in the tumor microenvironment in a mouse model [39]. The authors suggested that treatment with this nanoparticle reverted the immunosuppressive tumor microenvironment to a functionally active state that acts synergistically with chemotherapy for better anticancer response.

In addition, pH-responsive nanoparticles for delivery of PTX have been developed and are reported to improve cytotoxic activity compared to pH-insensitive nanoparticles [40]. In addition, the same study found that these particles can reverse the multidrug resistance of PTX-resistant cancer cells.

The most important determinant of therapeutic effect for PTX is transportation to the cancer cell. In the near feature, advanced technologies are likely to change the therapeutic strategies for the use of PTX in patients with UC.

Targeted Anti-Tumor Agents

Angiogenesis is a key step that primarily involves the generation of new blood vessels from the pre-existing vasculature. This step is essential for tumor growth and cancer cell dissemination from the primary tumor in almost all solid cancers. Proliferation, migration, and tube formation are the most important events of the angiogenic cascade. Partial or full inhibition of these events is thus crucial for anti-tumor activities in various malignancies, including UC. At present, many types of molecular-targeted therapy show anti-angiogenic activity.

PTX leads to cell death via the stabilization of microtubules active in malignant cells. In addition to such cytotoxic activity, PTX is an anti-angiogenic agent that inhibits endothelial cell proliferation, migration, and tube formation [17,41]. However, the main anti-tumoral effects of PTX seem to be achieved through the stabilization of microtubules. As a result, several targeted anti-angiogenic agents have been combined with PTX in trials for various malignancies [42,43].

Another interesting pharmacological characteristic of PTX is that the agent leads to increased levels of sunitinib. Heath et al. [42] reported maximum and total plasma exposures to sunitinib and its active metabolite were increased when given with PTX plus carboplatin in a phase I study of solid tumors. Unfortunately, this phenomenon led to increased risk of adverse events, but is interesting in terms of discussing new treatment strategies using PTX.

In recent years, the anti-angiogenic effects of dual drug-loaded polymeric micelles of PTX and rapamycin have been reported [44]. Rapamycin is an m-TOR inhibitor and causes cell arrest in the G1 phase and suppression of pro-angiogenic factors through down-regulation of hypoxia-mediated pathways [45,46]. Previous reports have shown that the anti-angiogenic activities of rapamycin differ from those of PTX [17,41]. This method takes advantage of each of the anti-tumoral and anti-angiogenic effects from combination therapy and the drug delivery system. In fact, the dual-loaded polymeric micelles exert synergistic anti-angiogenic activity and have strong potential for further development as an anti-angiogenic chemotherapy [44].

Table summarizes clinical trials using molecular-targeted therapy including PTX. The present state and near future of molecular-targeted therapy for UC have been reviewed by many investigators [47-49]. However, we and the authors of these reviews expect improvements in therapeutic efficacy and patient outcomes with the continued development of molecular-targeted therapies. Even so, innovations and the introduction of radical changes in treatment strategies and anti-cancer agents are necessary to fulfill these dreams.

Table. On-going clinical trials including paclitaxel for urothelial cancers

Combined drug/therapy	Objective	Phase	Start	Status	IDs of trials
Trastuzumab/ Radiation	MIBC after TUR	I/II	2005	Recruiting	NCT00238420
GEM and DOC /	Metastatic/ Unresectable	II	2007	Active	NCT00478361
Lapatinib/	Advanced	I	2006	Recruiting	NCT00313599
Pazopanib/	Refractory	II	2010	Recruiting	NCT01108055
Everolimus/	Advanced/ First	II	2010	Recruiting	NCT01215136
Abraxane/ Intravesical	Refractory	I/II	2007	Recruiting	NCT00583349
Abraxane Carbo GEM/	Neoadjuvant	II	2007	Active	NCT00585689
Abraxane Carbo GEM/	First -Advanced	II	2009	Active	NCT00995488
single/	Second-Advanced	II	2008	Recruiting	NCT00683059

Intravesical Treatment

As mentioned above, the main use of PTX has been for treating advanced bladder cancer, including muscle-invasive bladder cancer (MIBC) and/or metastatic diseases. On the other hand, approximately 60-70% of bladder cancer patients show non-MIBC (NMIBC) at diagnosis, and transurethral resection (TUR) is commonly performed as primary treatment [50]. In almost all NMIBC patients, radical resection is expected as appropriate TUR at that time. Unfortunately, even for bladder tumors with cancer cells limited to the mucosa, recurrences appear in about 70% of tumors and 30% progress to invasive tumors; the high malignant potential is linked to a high risk of subsequent metastasis and poor survival [51,52]. Prevention of recurrence into bladder mucosa is thus important for improving outcome and survival in patients with NMIBC. Intravesical therapy is a common method and several chemotherapeutic agents or some types of Bacillus Calmette-Guerin (BCG) are considered the best recourse, predominantly depending on stage of disease and presence of carcinoma in situ [53-55]. However, approximately a quarter to half of patients experience recurrence within 5 years after standard TUR and intravesical BCG therapy [53-55]. When such first line intravesical therapy fails, radical cystectomy is the option most likely to improve survival, because effective second line therapies have yet to be established. On the other hand, some patients refuse this life-altering method or are ineligible for the operation due to medical comorbidities or age. Development of novel anti-tumoral agents for intravesical therapy is thus necessary.

Several new anti-tumor agents for intravesical therapy have been studied by urologists and investigators. Comprehensive reviews and interesting investigations regarding strategies to enhance efficacy and current trends in intravesical therapy for NMIBC have been reported [56,57]. The present report thus takes note of intravesical therapy using taxanes.

In 1994, the in vitro activity and urine stability of taxanes, including PTX and docetaxel, were investigated and reported as effective against UC cell lines and stable in human urine for over 4 h [58]. These findings suggest that both PTX and docetaxel may be clinically useful agents for intravesical use against UC. The first human trial of intravesical taxane therapy was reported in 2006, showing an initial response rate for docetaxel of 56% [59]. In addition to this phase I trial, the efficacy and safety of docetaxel have been studied for the management of NMIBC refractory to standard intravesical therapies including BCG [60]. On the other hand, several reports have described 2 h intravesical administration of PTX as sufficient to produce anti-proliferative and apoptotic effects in 70-90% of human bladder tumors, and this agent represents a viable candidate drug for intravesical therapy according to a pharmacokinetic study [61,62]. Similar to docetaxel, PTX seems suitable and useful for intravesical therapy in NMIBC patients. However, to the best of our knowledge, no clinical trials of intravesical therapy have used pure PTX alone. Wosnitzer, et al. [63] used docetaxel and nano-particulate albumin-bound PTX, but not PTX, in a study on predictive markers for intravesical therapy in NMIBC. One major reason that PTX was not used for intravesical therapy was that Cremophor was used to solubilize the drug and, by trapping PTX in micelles, reduced the free fraction of PTX and consequently lowered drug penetration into bladder tissues [64]. On the other hand, the efficacy and safety of intravesical therapy using various PTX-loaded nanoparticles and PTX-based agents have been investigated in animal experiments and clinical studies

[63,65,66]. Those studies showed that effective administration of PTX to cancer cells leads to improvement of the anti-cancer effects against UC cells.

In recent years, a phase I trial of intravesical nanoparticle albumin-bound PTX (ABI-007) was performed for patients with BCG-refractory NMIBC [67]. In that trial, 18 patients received 6 weekly instillations of ABI-007 and no grade 2, 3, or 4 drug-related local toxicities were encountered. A phase II trial is thus being designed to evaluate the anti-tumoral effects of this regimen. In addition, the efficacy and safety of intravesical induction of PTX-hyaluronic acid or PTX gelatin nanoparticles have been investigated in clinical studies for NMIBC patients, including those with BCG-refractory cancer [68,69]. Thus, by improving delivery systems, PTX and its "family" may change treatment strategies, including for the prevention of tumor recurrence in patients with NMIBC.

Conclusion

Paclitaxel-based therapies are useful and promising for patients with UC. In special, development of delivery system, technology of nanoparticle and molecular targeted agents are speculated to benefit for prevention and therapy strategies in patients with UC. Although further studies and clinical trials are necessary for conclusion of these hypotheses, we emphasize that PTX and various drugs derived from PTX are one of key therapeutic agents in present and near future.

Acknowledgments

We are grateful to Mr. Takumi Shimogama and Shigeru Kanda for their outstanding support. This study was supported in part by a Grant-in-Aid from the Japan Society for the Promotion of Science (to Y.M.). However, the funder has no effect on study design, data collection and analysis, decision to publish, or preparation of the manuscript. No additional external funding was received for this study.

References

[1] Jemal, A., Siegel, R., Ward, E., Hao, Y., Xu, J. & Thun, M. J. (2009) Cancer Statistics, 2009. *CA Cancer J Clin*, *59*, 225 – 249.

[2] Vaughn, D. J. (2000). Paclitaxel and carboplatin in bladder cancer: recent development. *Eur J Cancer*, *36*, 7 – 12.

[3] Sternberg, C. N., Yagoda, A., Scher, H. I., Watson, R. C., Geller, N., Herr, H. W., Morse, M. J., Sogani, P. C., Vaughan, E. D. & Bander, N. et al. (1989). Methotrexate, vinblastine, doxorubicin, and cisplatin for advanced transitional cell carcinoma of the urothelium. *Cancer*, *64*, 2448 – 2458.

[4] Logothetis, C. J., Dexeus, F. H., Finn, L., Sella, A., Amato, R. J., Ayala, A. G. & Kilbourn, R. G. (1990). A prospective randomized trial comparing MVAC and CISCA chemotherapy for patients with metastatic urothelial tumors. *J Clin Oncol*, *8*, 1050 – 1055.

[5] Loehrer, P., Einhorn, L. H., Elson, P. J., Crawford, E. D., Kuebler, P., Tannock, I., Raghavan, D., Stuart-Harris, R., Sarosdy, M. F. & Lowe, B. A. et al. (1992). A randomized comparison of cisplatin alone or in combination with methotrexate, vinblastine and doxorubicin in patients with metastatic urothelial carcinoma of the bladder. A cooperative group study. *J Clin Oncol*, *10*, 1066 – 1073.

[6] Von der Maase, H., Hansen, S. W., Roberts, J. T., Dogliotti,. L., Oliver, T., Moore, M. J., Bodrogi, I., Albers, P., Knuth, A., Lippert, C. M., Kerbrat, P., Sanchez Rovira. P., Wersall, P., Cleall, S. P., Roychowdhury, D. F., Tomlin, I., Visseren-Grul, C. M. & Conte, P. F. (2000). Gemcitabine and cisplatin versus methotrexate, vinblastine, doxorubicin and cisplatin in ad advanced or metastatic bladder cancer: results of a large, randomized, multicenter phase III study. *J Clin Oncol*, *18*, 3068-3077.

[7] Culine, S. (2002). The present and future of combination chemotherapy in bladder cancer. *Semin Oncol*, *29*, 32 – 39.

[8] Cohen, M. H. & Rothmann, M. (2001). Gemcitabine and cisplatin for advanced, metastatic bladder cancer. *J Clin Oncol*, *19*, 1229-1231.

[9] Bellmunt, J., Ribas, A., Eres, N., Albanell, J., Almanza, C., Bermejo, B., Solé, L. A. & Baselga, J. (1997). Carboplatin-based versus cisplatin-based chemotherapy in the treatment of surgically incurable advanced bladder carcinoma. *Cancer*, *80*, 1966 – 1972.

[10] Galsky, M. D., Chen, G. J., Oh, W. K., Bellmunt, J., Roth, B. J., Petrioli, R., Dogliotti, L., Dreicer, R. & Sonpavde, G. (2012).

Comparative effectiveness of cisplatin-based and carboplatin-based chemotherapy for treatment of advanced urothelial carcinoma. *Ann Oncol*, *23*, 406 – 410.

[11] Waxman, J. & Barton, C. (1993). Carboplatin-based chemotherapy for bladder cancer. *Cancer Treat Rev*, *19*, Suppl C, 21 – 25.

[12] Saxman, S. B., Propert, K. J., Einhorn, L. H., Crawford, E. D., Tannock, I., Raghavan, D., Loehrer, P. J. Sr, & Trump, D. (1997). Long-term follow-up of a phase III intergroup study of cisplatin alone or in combination with methotrexate, vinblastine, and doxorubicin in patients with metastatic urothelial carcinoma: a cooperative group study. *J Clin Oncol*, *15*, 2564 – 2569.

[13] Bellmunt, J. & Albiol, S. (2007). Chemotherapy for metastatic or unresectable bladder cancer. *Semin Oncol*, *34*, 135-144.

[14] Bajorin, D. F. (2000). Paclitaxel in the treatment of advanced urothelial cancer. *Oncology* (Williston Park), *14*, 43 – 52.

[15] Demidenko, Z. N., Kalurupalle, S., Hanko, C., Lim, C. U., Broude, E. & Blagosklonny, M. V. (2008). Mechanism of G1-like arrest by low concentrations of paclitaxel: next cell cycle p53-dependent arrest with sub G1 DNA content mediated by prolonged mitosis. *Oncogene*, *27*, 4402-4410.

[16] Tso, P. H., Morris, C. J., Yung, L. Y., Ip, N. Y. & Wong, Y. H. (2009). Multiple Gi proteins participate in nerve growth factor-induced activation of c-Jun N-terminal kinases in PC12 cells. *Neurochem Res*, *34*, 1101-1112.

[17] Belotti, D., Vergani, V., Drudis, T., Borsotti, P., Pitelli, M. R., Viale, G., Giavazzi, R. & Taraboletti, G. (1996). The microtubules-affecting drug paclitaxel has antiangiogenic activity. *Clin Cancer Res*, *2*, 1843 – 1849.

[18] Rowinsky, E. K., Cazenave, L. A. & Donehower, R. C. (1990). Taxol: a novel investigational antimicrotube agent. *J Natl Cancer Inst*, *82*, 1247-1259.

[19] Stanton, R. A., Gernert, K. M., Nettles, J. H. & Aneja R. (2011). Drugs that target dynamic microtubules: a new molecular perspective. *Med Res Rev*, *31*, 443-448.

[20] Papamichael, D., Gallagher, C. J., Oliver, R. T., Johnson, P. W. & Waxman, J. (1997). Phase II study of paclitaxel in pretreated patients with locally advanced/metastatic cancer of the bladder and ureter. *Br J Cancer*, *75*, 606 – 607.

[21] Raghavan, D. (2003). progress in the chemotherapy of metastasic cancer of the urinary tract. *Cancer*, *97* (8 supple), 2050 – 2055.

[22] Shord, S. S., Faucette, S. R., Gillenwater, H. H., Pescatore, S. L., Hawke, R. L., Socinski, M. A. & Lindley, C. (2003). Gemcitabine pharmacokinetics and interaction with paclitaxel in patients with advanced non-small-cell lung cancer. *Cancer Chemother Pharmacol*, *51*, 328 – 336.

[23] Niegisch, G., Fimmers, R., Siener, R., Park, S. I., Albers, P. (2011). German Association of Urological Oncology Bladder Cancer Group. Prognostic factors in second-line treatment of urothelial cancers with gemcitabine and paclitaxel (German Association of Urological Oncology trial AB20/99). *Eur Urol*, *60*, 1087-1196.

[24] Ikeda, M., Matsumoto, K., Tabata, K., Minamida, S., Fujita, T., Satoh, T., Iwamura, M. & Baba, S. (2011). Combination of gemcitabine and paclitaxel is a favorable option for patients with advanced or metastatic urothelial carcinoma previously treated with cisplatin-based chemotherapy. *Jpn J Clin Oncol*, *41*, 1214-1220.

[25] Tsavaris, N., Kosmas, C., Skopelitis, H., Dimitrakopoulos, A., Kopterides, P., Bougas, D., Stravodimos, K., Mitropoulos, D., Alamanis, C. & Giannopoulos, A. (2005). Methotrexate-paclitaxel-epirubicin-carboplatin (M-TEC) combination chemotherapy in patients with advanced bladder cancer: an open label phase II study. *J Chemother*, *17*, 441-448

[26] Halim, A. & Abotouk, N. (2012). methotrexate^paclitaxel-epirubicin-carboplatin as second line chemotherapy in patients with metastatic transitional cell carcinoma of the bladder pretreated with cisplatin-gemcitabine: a phase II study. *Asia-Pac J Clin Oncol*, *9*, 60 – 65.

[27] Kattan, J. G., Boutros, C. Y., Farhat, F. S., Chahine, G. Y., Musallam, K. M. & Ghosn, M. G. (2012). Sequential therapy with gemcitabine and carboplatin followed by paclitaxel as first line treatment for advanced urothelial cancer. *J Cancer*, *3*, 362 – 368.

[28] Singla, A. K., Garg, A. & Aggarwal, D. (2002) Paclitaxel and its formations. *Int J pharm*, *235*, 179 – 192.

[29] Kloover, J. S., den Bakker, M. A., Gelderblom, H. & van Meerbeeck, J. P. Fatal outcome of a hypersensitivity reaction to paclitaxel: a critical review of premedication regimens. *Br J Cancer*, *90*, 304-305.

[30] Ganning, A.E., Brunk, U. & Dallner, G. (1984), Phthalate esters and their effect on the liver. *Hepatology*, *4*, 541-547.

[31] Kim, S. C., Yoon, H. J., Lee, J. W., Yu, J., Park, E. S. & Chi, S. C. (2005) Investigation of the release behavior of DEHP from infusion sets by paclitaxel-loaded polymeric micelles. *Int J Pharm*, *293*, 303-310.

[32] Sparreboom, A., Scripture, C. D., trieu, V., Williams, P. J., De, T., Yang, A., Beals, B., Figg, W. D., Hawkins, M. & Desai, N. (2005). Comparative preclinical and clinical pharmacokinetics of a cremophor-free, nanoparticle albumin-bound paclitaxel (ABI-007) and paclitaxel formylated in Cremophor (Taxol). *Clin Cancer Res*, *11*, 4136- 4143.

[33] Gradishar, W. J., Tjuladin, S., Davidson, N., Shaw, H., Desai, N., Bhar, P., Hawkins, M. & O'Shaughnessy, J. (2005). Phase III trial of nanoparticle albumin-bound paclitaxel compared with polyethylated catrol oil-based paclitaxel in women with breast cancer. *J Clin Oncol*, *23*, 7794-7803.

[34] Zhang, C., Awasthi, N., Schwarz, M. A., Hinz, S. & Schwarz, R. E. (2013). Superior antitumor activity of nanoparticle albumin-bound paclitaxel in experimental gastric cancer. *PLOS ONE*, *8*, e58037-???

[35] Fu, S., Naing, A., Moulder, S. L., Culotta, K. S., Madoff, D. C., Ng, C. S., Madden, T. L., Falchook, G. S., Hong, D. S. & Kurzrock, R. (2011). Phase I trial of hepatic arterial infusion of nanoparticle albumin-bound paclitaxel: toxicity, pharmacokinetics, and activity. *Mol Cancer Ther*, *10*, 1300-1307.

[36] Tsimberidou, A. M., Ye, Y., Wheler, J., Naing, A., Hong, D., Nwosu, U., Hess, K. R. & Wolff, R. A. (2013). A phase I study of hepatic arterial infusion of nab-paclitaxel in combination with intravenous gemcitabine and bevacizumab for patients with advanced cancers and predominant liver metastases. *Cancer Chemother Pharmacol*, *71*, 955-963.

[37] Chun, C., Lee, S. M., Kim, S. Y., Yang, H. K. & Song, S. C. (2009). Thermosensitive poly(organophosphazene)-paclitaxel conjugate gels for antitumor applications. *Biomaterials*, *30*, 2349-2360.

[38] Kim, J. H., Lee, J. H., Kim, K. S., Na, K., Song, S. C., Lee, J. & Kuh, H. J. (2013). Intratumoral delivery of paclitaxel using a thermosensitive hydrogel in human tumor xenografts. *Arch Pharm Res*, *36*, 94-101.

[39] Roy, A., Singh, M. S., Upadhyay, P. & Bhaskar, S. Nanoparticle mediated co-delivery of paclitaxel and a TLR-4 agonist results in tumor regression and enhanced immune response in the tumor microenvironment of a mouse model. *Int J Pharm*, *445*, 171-180.

[40] He, H., Chen, S., Zhou, J., Dou, Y., Song, L., Che, L., Zhou, X., Chen, X., Jia, Y., Zhang, J., Li, S. & Li, X. (2013). Cyclodextrin-derived pH-responsive nanoparticles for delivery of paclitaxel. *Biometrials*, in press.

[41] Ng, S. S., Figg, W. D. & Sparreboom, A. Taxane-mediated antiangiogenesis in vitro: influence of formation vehicles and binding proteins. *Cancer Res*, *64*, 821 – 824.

[42] Heath, E. I., Blumenschein, Jr., G. R., Cohen, R. B., LuRosso, P. M., LoConte, N. K., Kim, S. T., Ruiz-Garcia, A., Chao, R. C. & Wilding G. (2011). Sunitinib in combination with paclitaxel plus carboplatin in patients with advanced solid tumors: phase I study results. *Cancer Chemother Pharmacol*, *68*, 703 – 712.

[43] Marchion, D. C., Bicaku, E., Xiong Y., Zgheib. N. B., Al Sawah, E., Stickles, X. B., Judson, P. L., Lopez, A. S., Cubitt, C. L., Gonzalez-Bosquet, J., Wenham, R. M., Apte, S. M., Berglund, A. & Lancaster, J. M. (2013). A novel c-Met inhibitor, MK8033, synergizes with carboplatin plus paclitaxel to inhibit ovarian cancer cell growth. *Oncol Rep*, in press.

[44] Mishra, G. P., Nguyen, D. & Alani, W. G. (2013). Inhibitory effect of paclitaxel and rapamycin individual and dual drug-loaded polymeric micelles in the angiogenic cascade. *Mol Pharmaceutics*, in press.

[45] Faivre, S., Kroemer, G. & Raymond, E. (2006). Current development of mTOR inhibitors as anticancer agents. *Nat rev Drug Discovery*, *5*, 671 – 688.

[46] Guertin, D. A. & Sabatini, D. M. (2007). Defining the role of m-TOR in cancer. *Cancer Cell*, *12*, 9 – 22.

[47] Bellmunt, J., Albiol, S., Suárez, C. & Albanell, J. (2009). Optimizing therapeutic strategies in advanced bladder cancer: update on chemotherapy and the role of targeted agents. *Crit Rev Oncol Hematol*, *69*, 211-222.

[48] Serrano, C., Morales, R., Suárez, C., Núñez, I., Valverde, C., Rodón, J., Humbert, J., Padrós, O. & Carles, J. (2012). Emerging therapies for urothelial cancer. *Cancer Treat Rev*, *38*, 311-317.

[49] Mitsui, Y., Yasumoto, H., Arichi, N., Honda, S., Shiina, H. & Igawa, M. (2012). Current chemotherapeutic strategies against bladder cancer. *Int Urol Nephrol*, *44*, 431-441.

[50] Hussain, Sa. & James, N. D. (2003). The systemic treatment of advanced and metastatic bladder cancer. *Lancet Oncol*, *4*, 489-497.

[51] Stein, J. P., Grossfeld, G. D., Ginsberg, D. A., Esrig, D., Freeman, J. A., Figueroa, A. J., Skinner, D. G. & Cote, R. J. (1998). Prognostic markers in bladder cancer: a contemporary review of the literature. *J Urol*, *160*, 645-659.

[52] Kurth, K. H., Bouffioux, C., Sylvester, R., van der Meijden, A. P., Oosterlinck, W. & Brausi, M. (2000). Treatment of superficial bladder tumors: achievements and needs. The EORTC Genitourinary Group. *Eur Urol*, *37*, 1-9.

[53] Morales, A. (1984). Long-term results and complications of intracavitary bacillus Calmette-Guerin therapy for bladder cancer. *J Urol*, *132*, 457 – 459.

[54] Herr, H. W., Wartinger, D. D., Fair, W. R. & Oettgen H. F. (1992). Bacillus Calmette-Guerin therapy for superficial bladder cancer: a 10-year followup. *J Urol*, *147*, 1020 – 1023.

[55] Pansadoro, V., Emiliozzi, P., Defidio, L. Donadio, D., Florio, A., Maurelli, S., Lauretti, S. & Sternberg, C. N. (1995). Bacillus Calmette-Guerin in the treatment of stage T1 grade 3 transitional cell carcinoma of the bladder: long-term results. *J Urol*, *154*, 2054 – 2058.

[56] Smaldone, M. C., Gayed, B. A., Tomaszewski, J. J. & Gingrich, J. R. (2009). Strategies to enhance the efficacy of intravescical therapy for non-muscle invasive bladder cancer. *Minerva Urol Nefrol.*, *61*, 71 – 89.

[57] Nargund, V. H., Tanabalan, C. K. & Kabir, M. N. (2012). Management of non-muscle-invasive (superficial) bladder cancer. *Semin Oncol*, *39*, 559 – 572.

[58] Rangel, C., Niell, H., Miller, A. & Cox, C. (1994). Taxol and taxotere in bladder cancer: in vitro activity and urine stability. *Cancer Chemother Pharmacol*, *33*, 460 – 464.

[59] McKiernan, J. M., Masson, P., Murphy, A. M. Goetzl, M., Olsson, C. A., Petrylak, D. P., Desai, M. & Benson, M. C. (2006). Phase I trials of intravesical docetaxel in the management of superficial bladder cancer refractory to standard intravesical therapy. *J Clin Oncol*, *24*, 3075 – 3080.

[60] Barlow, L. J., McKiernan, J. M. & Benson, M. C. (2009). The novel use of intravesical docetaxel for the treatment of non-muscle invasive bladder cancer refractory to BCG therapy: a single institution experience. *World J Urol*, *27*, 331 – 335.

[61] Au, J. L., Kalns, J., Gan, Y. & Wientjes, M. G. (1997). Pharmacologic effects of paclitaxel in human bladder tumors. *Cancer Chemother Pharmacol*, *41*, 69 – 74.

[62] Song, D., Wientjes, M. G. & Au, J. L. (1997). Bladder tissue pharmacokinetics of intravesical taxol. *Cancer Chemother Pharmacol*, *40*, 285 – 292.

[63] Wosnitzer, M. S., Domingo-Domenech, J., Castillo-Martin, M., Ritch, C., Mansukhani, M., Petrylack, D. P., Benson, M. C., McKiernan, J. M. & Cordon-Cardo, C. (2011). Predictive value of microtubule associated proteins tau and stathmin in patients with nonmuscle invasive bladder cancer receiving adjuvant intravesical taxane therapy. *J Urol, 186,* 2094-2100.

[64] Knemeyer, I., Wientjes, M. G., & Au, J. L-S. (1999). Cremophor reduces paclitaxel penetration into bladder wall during intravesical treatment. *Cancer Chemother Pharmacol, 44,* 241 – 248.

[65] Lu, Z., Yeh, T. K., Tsai, M., Au, J. L. & Wientjes, M. G. (2004). Paclitaxel-loaded gelatin nanoparticles for intravesical bladder cancer therapy. *Clin Cancer Res, 10,* 7677 – 7684.

[66] Rosato, A., Banzato, A., De Luca, G., Renier, D., Bettella, F., Pagano, C., Esposito, G., Zanovello, P. & Bassi, P. (2006). HYTAD1-p20: a new paclitaxel-hyaluronic acid hydrosoluble bioconjugate for treatment of superficial bladder cancer. *Urol Oncol, 24,* 207-215.

[67] McKiernan, J. M., Barlow, L. J., Laudano, M. A., Mann, M. J., Petrylak, D. P. & Benson, M. C. (2011). A phase I trials of intravesical nanoparticle albumin-bound paclitaxel in the treatment of bacillus Calmette-Guérin refractory nonmuscle invasive bladder cancer. *J Urol, 186,* 448 – 451.

[68] Lu, Z., Yeh, T. K., Wang, J., Chen, L., Lyness, G., Xin, Y., Wientjes, M. G., Bergdall, V., Couto, G., Alvarez-Berger, F., Kosarek, C. E. & Au, J. L. (2011). Paclitaxel gelatin nanoparticles for intravesical bladder cancer therapy. *J Urol, 185,* 1478-1483.

[69] Bassi, P. F., Volpe, A., D'Agostino, D., Palermo, G., Renier, D., Franchini, S., Rosato, A. & Racioppi, M. (2011). Paclitaxel-hyaluronic acid for intravesical therapy of bacillus Calmette-Guérin refractory carcinoma in situ of the bladder: results of a phase I study. *J Urol 185,* 445-449.

In: Paclitaxel
Editor: Diego Morales
ISBN: 978-1-62808-549-5

Chapter 5

Mesenchymal Stromal Cells Used for Carrying and Delivering Paclitaxel

Augusto Pessina[1,*], Arianna Bonomi[1], Emilio Ciusani[2], Luisa Pascucci[3], Eugenio Parati[2] and Giulio Alessandri[2]

[1]Department of Biomedical, Surgical and Dental Sciences, University of Milan, Italy

[2]Laboratory of Clinical Pathology and Neurogenetic Medicine; Fondazione IRCCS, Neurological Institute Carlo Besta, Milan, Italy

[3]Department of Biopathological Science and Hygiene of Animal and Food Productions - Section of Veterinary Anatomy, University of Perugia, Italy

[4]Cellular Neurobiology Laboratory, Department of Cerebrovascular Diseases; Fondazione IRCCS, Neurological Institute Carlo Besta, Milan, Italy

Abstract

In recent years many strategies have been developed to obtain high concentrations of anti-cancer drugs effective on tumor mass. Genetically

[*] Corresponding author: augusto.pessina@unimi.it.

modified Mesenchymal Stromal Cells (MSCs) have been indicated as a candidate to deliver such drugs.

In a previous study, we found that mouse-bone-marrow-derived MSCs (BM-MSCs) exposed to high doses of Doxorubicin were able to incorporate and release the drug, acquiring an anti-proliferative potential towards adjacent co-cultured haematopoietic stem cells. We thus hypothesized whether freshly isolated human BM-MSCs, adipose tissue MSCs (AT-MSCs) and skin-derived stromal fibroblasts (hSDFs) primed *in vitro* with Paclitaxel (PTX) may acquire anti-tumor activity when localized near cancer cells.

Incorporation of PTX into MSCs was studied by using FITC-labelled-PTX and analyzed by FACS and confocal microscopy. Release of PTX in culture medium by PTX primed BM-MSCs (BM-MSCsPTX) was investigated by HPLC. We showed that BM-MSCs, AT-MSCs and hSDFs acquired strong anti-tumor capacity, once primed with PTX. Cells were able to uptake PTX and release it in a time-dependent manner and the released drug was active *in vitro* against proliferation of leukemia, osteosarcoma, prostatic carcinoma and neuroblastoma cell lines.

In vitro, both BM-MSCs and BM-MSCsPTX bound and formed rosettes with leukemia cells (LCs). Ultrastructural analysis of rosettes did not show morphological alterations of BM-MSCsPTX, while the attached LCs were apoptotic and necrotic. BM-MSCs and BM-MSCsPTX released molecules that reduced LCs adhesion to microvascular endothelium (HMECs) and down-modulated ICAM-1 and VCAM-1 on HMECs. *In vivo* studies showed that BM-MSCsPTX had a strong anti-tumor activity if co-injected with tumor cells or injected intra-tumor. Established subcutaneous tumor nodules in mice treated with BM-MSCsPTX showed a significant size reduction as well as decreased vascularisation. Priming human MSCs with PTX is a simple procedure that does not require any genetic cell manipulation. The harvesting of BM-MSCs, AT-MSCs and hSDFs do not require invasive procedures and is without any particular pain or discomfort to donor patients. Our studies demonstrate for the first time that MSCs isolated from different tissue sources can uptake and subsequently slowly release PTX. This suggests that MSCs could become a tool for drug delivery due to its ability to increase efficacy and target selectivity in cancer therapy.

Introduction

Multipotent mesenchymal stromal cells (MSCs) are non hematopoietic cells of mesodermal origin that are present in several adult tissues and organs: brain, liver, kidney, lung, bone marrow, muscle, thymus, pancreas, skin, adipose tissue, umbilical cord, Wharton's jelly, placenta and deciduous teeth [1-5]. The highest degree of lineage plasticity has been imputed to bone marrow derived MSCs (BM-MSCs). These cells are easily handled *in vitro* and to date, most pre-clinical studies have been performed with them [6]. The harvesting of bone marrow needs an invasive procedure which yields a small number of cells that require *in vitro* manipulation to be expanded. However, adipose tissue (AT) is an alternate, abundant and easily accessible MSCs source which has received attention in recent years. AT-MSCs possess expansion potential, differentiation capacity and immunophenotype nearly identical to BM-MSCs [7]. Moreover, fibroblasts, considered mature MSCs and particularly abundant in the connective of each organ and tissue, can be easily isolated from skin biopsies without any discomfort for the donor. Fibroblasts share several phenotypic and functional features of immature MSCs and sometimes they are not distinguishable from BM or AT-MSCs [8,9].

The main goal in cancer chemotherapy is to localize the drug effect in the tumor microenvironment in order to kill as many tumor cells (TCs) as possible while producing the lowest collateral toxicity. To achieve this purpose, different approaches have been proposed, such as toxic immunoconjugates [10], nanoparticles [11], manipulated stem cells [12] and, more recently, engineered mesenchymal stromal cells (MSCs). This last approach seems to represent a good choice to deliver anti-tumor agents due to the easy adaptability of MSCs to culture conditions necessary for *in vitro* manipulation and their capacity for homing to pathological tissues when injected *in vivo*. MSCs have been genetically modified to over-express several anti-tumor factors, such as interleukins, interferons, pro-drugs, oncolytic viruses, anti-angiogenic agents, pro-apoptotic proteins and growth factors antagonists able to kill TCs both *in vitro* and *in vivo* [6]. Despite the promising results on animal models, many scientists/scientific and regulatory agencies do not consider the genetic manipulation of MSCs for clinical application in humans free from risks [13]. Furthermore, the use of engineered MSCs secreting toxic compounds pose a particular problem regarding the cells' fate after the eradication of the targeted malignant process [14].

Bone marrow stromal cells exhibited drug metabolizing activity and were also able to affect hematopoietic precursors through the intracellular accumulation of Doxorubicin (DXR) and its delayed release [15]. These data suggested that BM stromal cells may have some role in modulating drug toxicity on hematopoietic progenitors. As reported by Pessina et al. [16] murine stromal cells cultured in the presence of DXR were able to uptake significant amounts of the drug without showing significant signs of toxicity. By contrast, hematopoietic stem cells (HSCs) from BM were very sensitive to DXR. Interestingly, if HSCs were co-cultured with BM stromal cells first primed with DXR, a significant inhibition of HSCs-induced colony formation units (CFU) was observed. This observation indicated that murine stromal cells may act as a reservoir for DXR that, subsequently, may release some DXR metabolites or even DXR in its original form, leading to HSCs-induced CFU inhibition.

Based on the above reported data on murine stromal cells properties, it has been investigated whether this behaviour was restricted to these cells only or was also a capacity of other stromal cells. At first, the studies were conducted on human BM derived MSCs by priming *in vitro* these cells with Paclitaxel (PTX), an anticancer agent which also demonstrates an important anti-angiogenic activity [17, 18]. PTX is a highly lipophilic drug (derived from Taxus brevifolia) which is very active against many solid tumors and endothelial cell proliferation. The drug induces the mitotic arrest of the cell by promoting microtubule polymerization [19, 20] without interfering with other cell functions. In further studies, our investigations have been expanded to human AT-MSCs obtained from omentum fat samples [21] and to human skin-derived fibroblasts (hSDFs) [22]. These last ones, similar to mesenchymal stromal cells, can be recruited by growing tumors and, for this feature, they have been proposed for use in cell therapy [23, 24]. This chapter summarizes the results obtained by our group on the application of MSCs as a carrier of Paclitaxel. We were able to demonstrate that *in vitro* priming of MSCs (from bone marrow and adipose tissue) and hSDFs with a very high dose of PTX may represent an interesting tool to load cells with the drug which can then be delivered *in vivo* directly into a tumor microenvironment. This property of stromal cells represents an interesting system that could contribute to new therapeutic applications using MSCs as carrier for drug delivery.

Priming of BM-MSCs with Paclitaxel

Cultures of human BM-MSCs were expanded and characterized. The BM-MSCs analysis confirmed the expression of MSCs markers (positivity for CD44, CD73, CD90, CD105, and HLA-I and negative for CD14, CD31, CD34, CD45, CD80 and HLA-II). When cultured under differentiating conditions, they acquired osteo-adipo and chondroblasts markers [17, 25].

In a preliminary MTT assay, BM-MSCs sensitivity to PTX was evaluated in a 7-days anti-proliferation assay and in a 24 hours cytotoxicity test. BM-MSCs were sensitive to the anti-proliferative activity of PTX, according to a dose-dependent kinetics with IC_{50} values of 4.07 ± 1.75 ng/ml; by contrast, BM-MSCs were strongly resistant to PTX direct cytotoxicity, with a 15% cell death even at concentrations higher than 10,000 ng/ml.

Based on these data, sub-confluent cultures of BM-MSCs were incubated at 37°C in air + 5% CO_2 for 24 hours with 2,000 ng/ml PTX (enough to completely block cell proliferation, but unable to affect cell viability). At the end of the incubation, the cells were trypsinized, washed twice and seeded in a new flask.

The effect on cell proliferation exerted by the PTX priming process of BM-MSCs was evaluated by cell cycle analysis that showed significant accumulation of cells in S and, in a minor degree, G2/M phase. BM-MSCs viability was not significantly affected and cells detached, washed and sub-cultured in the absence of drug gave a cell monolayer with a cell viability in the range of controls and a cell cycle pattern completely restored after 72 h. These data have been confirmed by the percentages of apoptotic/ necrotic cells counted in the different experimental conditions by Annexin assay and are in agreement with the reports of other authors on the sensitivity of stromal cells to Paclitaxel [26, 27].

PTX Internalization into BM-MSCs

The internalization of PTX into BM-MSCs was investigated over time by confocal microscopy using fluorescent PTX (PTX-F). Although the internalization was appreciable already after one hour of priming, after 24 hours (end of priming) the staining was more intense and localized inside vesicles near the cell membrane suggesting a possible secretion. To investigate the origin of vesicles, a co-localization study was performed in BM-MSCs by using PTX-F and BODIPY® TR ceramide, a marker specific for Golgi apparatus and Golgi derived vesicles. We observed a high level of co-localization, meaning that PTX-F was internalized inside Golgi-derived vesicles. Thus, PTX-F accumulates in vesicles instead of accumulating in microtubules [19], just as many other xenobiotics do [28]. To investigate if PTX priming induced alterations on BM-MSCs, a fine morphological analysis by transmission electron microscopy (TEM) was performed. No significant changes were observed in PTX primed cells when compared to untreated BM-

MSCs, except for the presence of numerous membrane-bounded vacuoles [18] (Figure 1).

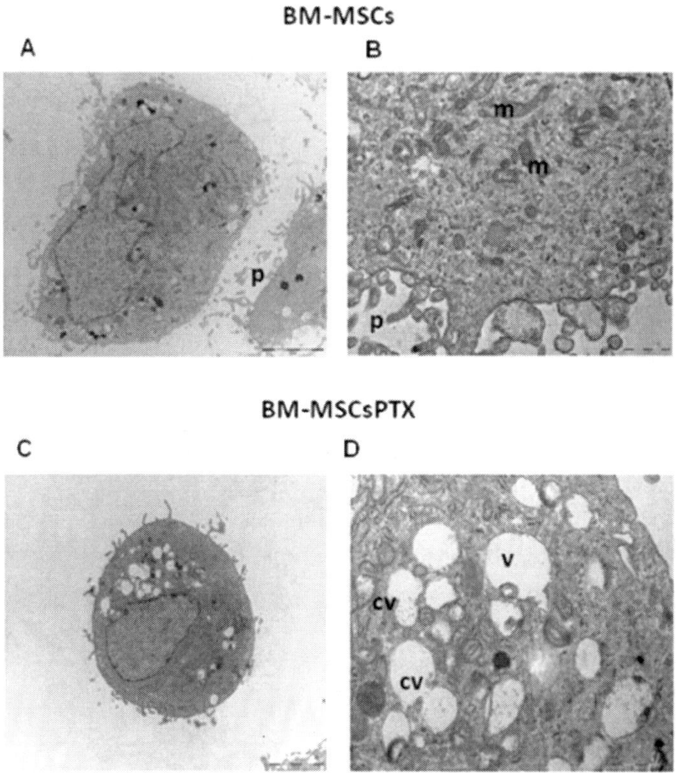

p = pseudopodal processes, m = mitochondria, v = membrane bounded vacuoles, cv = confluent vacuoles.

Figure 1. Transmission Electron Microscopy of BM-MSCs and BM-MSCsPTX.

PTX Uptake and Release by BM-MSCs

At the end of priming, BM-MSCs were trypsinized, washed and subcultured. 24 hours later, their conditioned media (CM) were collected and tested *in vitro* for their anti-tumoral activity on a reference leukemia cell line (MOLT-4) very sensitive to PTX (IC_{50} of 1.48 ± 1.06 ng/ml) [29]. The CM from cultures of BM-MSCsPTX (BM-MSCsPTX-CM) produced a strong dose-dependent anti-proliferative effect on MOLT-4, equivalent to those obtained with pure PTX at doses from 0.39 to 50 ng/ml. By contrast, the CM from untreated cells were uneffective. Comparing the inhibitory activity of pure PTX and CM on MOLT-4, we calculated the PTX equivalent concentration (PEC) in the CM used to estimate the PEC released by a single cell (PEC picograms/cell). The antitumoral activity of CM from PTX-treated cells was expressed as PTX equivalent concentration (PEC) according to the following algorithm: PEC (ng/ml) = $DF_{50}CM$ x $IC_{50}PTX$. $DF_{50}CM$ is the dilution factor at which the 50% of inhibition was observed with the CM and $IC_{50}PTX$ the concentration (ng/ml) of pure PTX producing 50% of inhibition. The PTX release at 24 hours expressed as PEC was 1.01±0.4 pg/cell. Primed cells frozen and thawed after three months were again able to release the drug after culture. The presence of PTX and the absence of the main PTX metabolites in BM-MSCsPTX-CM were confirmed by HPLC analysis. By replacing and collecting the CM of BM-MSCsPTX at different intervals of time, we calculated the time- dependent release of PEC. Detectable activity of PEC was present in the CM after 2 h of culture of BM-MSCsPTX, reaching a PEC of 1 pg/cell during the first 24 h of culture, and a maximum concentration of about 1.7-2.0 pg/cell at 144 h. Since these values did not increase with longer incubation, we estimated that around 25-30% of the total PEC found in cell lysate was retained by the cells and never released.

The mechanism of PTX uptake by cells and its binding to microtubules has been extensively studied [28, 30] but little is known about the cells capacity to release PTX or about the chemotherapy resistance of MSCs [26,31]. A report from Chaudhary and Roninson [32] demonstrated that multidrug efflux pump in BM derived HSCs involved P-glycoprotein (P-gp), a mechanism claimed for multidrug resistance (MDR) in cancer cells [32, 33]. P-gp is very sensitive to Verapamil (VP), an important specific first-generation efflux pump inhibitor [34]. In our investigations, we observed that BM-MSCs expressed very high levels of P-gp but, because their PTX sensitivity was not very much affected by VP treatment, we can not exclude that, besides P-gp, other mechanisms of MDR may be involved. An important role can be also exerted by exosomes and microvesicles. Exosomes are the most extensively characterized class of secreted membrane vesicles that carry proteins and RNAs for intercellular communication. There is evidence that cells can communicate through the transfer of vesicles and particles carrying lipids, proteins, mRNAs, miRNAs and DNAs. It could be of interest to verify if microvesicles are involved also in physiological mechanism to excrete PTX [35-38].

In Vitro Anti Tumor Proliferation Activity of BM-MSCsPTX

The *in vitro* anti-tumor potential of BM-MSCsPTX and the effect of BM-MSCsPTX-CM were evaluated on two human cell lines: glioblastoma T98G and prostate carcinoma DU145. The addition of BM-MSCsPTX-CM at 1:4 to 1:8 dilution induced up to 80% of growth inhibition on all tumor cell lines. Combining these data with those from the kinetics release of PTX tested on standard cell model (MOLT-4 leukemia cells), we estimated that in 24 hours, 3 x 10^5 BM-MSCsPTX may release around 50-60 ng/ml of PEC which corresponds to concentrations 3-5 fold the IC_{90} values of PTX for the cancer cells studied. The addition of control BM-MSCs-CM did not affect TCs proliferation.

The direct capacity of BM-MSCsPTX to inhibit the proliferation of different tumor cell lines was confirmed by a co-culture assay in which BM-MSCsPTX and tumor cells were mixed at different ratios (1:1, 1:10 and 1:100). The presence of BM-MSCsPTX inhibited proliferation of all tumor cell lines tested, according to their proportional presence. The addition of control BM-MSCs did not substantially affect tumor proliferation (just a low significant growth inhibition was observed at 1:1 ratio for all the tumor cell lines tested).

In Vitro Anti-Angiogenic Properties of BM-MSCsPTX

Because PTX is able to inhibit angiogenesis [20, 39], we investigated whether BM-MSCsPTX could affect angiogenesis *in vitro* in comparison to the pure drug. BM-MSCsPTX was tested both on macrovascular (HUVECs, human umbilical vein endothelial cells) and on microvascular endothelial cells (HMECs) proliferation. A ratio 1:1 and 1:5 of BM-MSCsPTX/ECs produced a strong cytotoxic effect on both HUVECs and HMECs. At ratio 1:10, no cytotoxicity but significant ECs growth inhibition was observed. At ratio 1:5, BM-MSCsPTX initiated killing HMECs during the first 24 hours of co-culture and, after 72 hours of incubation, very few HMECs survived. Control BM-MSCs did not affect ECs proliferation.

The anti-angiogenic potential of BM-MSCsPTX was also investigated by using the rat aorta ring assay [40] in which angiogenesis was quantified by scoring the number of microvessels arising from aorta rings. The addition of 1:2 and 1:4 dilutions of BM-MSCsPTX-CM greatly reduced spontaneous or VEGFa-induced sprouting of neocapillaries from aorta rings. 1:8 dilution of BM-MSCsPTX-CM, which inhibited ECs proliferation, did not affect aorta ring capillaries formation. BM-MSCsPTX-CM was also able to induce capillary regression if added to aorta rings after 7 days of culture and produced several zones of vessel necrosis. The addition of BM-MSCs-CM did not substantially affect capillary formation.

Leukemia Cells Binding and Damaging by BM-MSCs

In BM-MSCs and MOLT-4 leukemia cells (LCs) co-colture assay, we observed that LCs were preferentially anchored on BM-MSCs adherent to plastic. These results seems to suggest that BM-MSCs have the capacity to attract and bind LCs and the PTX priming does not alter these properties. To confirm the capacity of BM-MSCs to bind LCs, a rosette test was performed, by mixing LCs and BM-MSCs (control or PTX primed): they were able to give rise to structures formed by a central nucleus of a few aggregated BM-MSCs surrounded by a mono or multiple layers crown of LCs. Also BM-MSCsPTX were able to form rosette but LCs appeared damaged, necrotic and /or apoptotic: electron microscopy analysis showed marked changes in LCs, both nuclear and cytoplasmic, after the interaction with BM-MSCsPTX (Figure 2).

To better investigate the functional mechanism by which BM-MSCs attract and bind LCs, a transwell assay was used to evaluate their chemotactic activity. The presence of BM-MSCs-CM in the lower well produced a significant increment of time-dependent migration of LCs seeded on the upper well. The evaluation of the interaction between LCs and BM-MSCs by scanning electron microscopy suggests a role for extracellular matrix proteins in mediating BM-MSCs-LCs binding because no apparent specialization of the plasma membrane interaction between the two cells was observed.

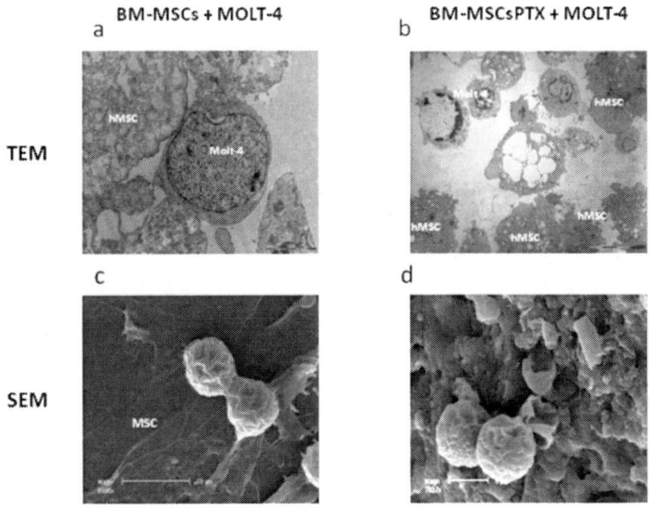

Figure 2. Ultrastructural analysis by Transmission (TEM) and Scanning (SEM) Electron Microscopy of BM-MSCs and MOLT-4 rosette.

In Vivo Inhibition of Tumor Growth by BM-MSCsPTX

In vitro data demonstrated that BM-MSCsPTX were strongly inhibitory for tumor cells (TCs) and even cytotoxic for endothelial cells. To see whether they may affect tumor growth *in vivo*, BM-MSCsPTX and control BM-MSCs were co-injected at ratio 1:5 with either human DU145 and MOLT-4 or mouse B16 melanoma cancer cells in immunodeficient mice. Control groups were injected with TCs alone or TCs mixed with 2,000 ng/ml of free PTX. BM-MSCsPTX mixed with MOLT-4 completely blocked tumor takes; the co-injection with DU145 produced a significant delay in tumor takes and reduced both DU145 and B16 tumor growth. The effect on the DU145 tumor progression exerted by BM-MSCsPTX was greater than that on B16. Tumor weights of DU145 and B16 were significantly reduced by the presence of BM-MSCsPTX and 25% of mice treated with DU145 and BM-MSCsPTX were also tumor free. Interestingly, the co-injection of tumor cells with a high concentration of free PTX did not delay tumor takes or affect growth which suggests that time-dependent release of PTX by BM-MSCsPTX is necessary to affect cancer cells proliferation.

We investigated also if BM-MSCsPTX could inhibit the growth of an established tumor nodule, by implanting MOLT-4 subcutaneously in nude mice until forming a nodule of approximately 10 mm in diameter. The intra tumor treatment was repeated every 7 days for a total of 4. We observed that only the treatment by BM-MSCsPTX greatly reduced tumor volume and weight while BM-MSCs or pure PTX did not exert any significant reduction. Immunohistochemical studies of MOLT-4 tumor specimens derived from mice treated with BM-MCSsPTX showed a significant reduction of tumor vascularization; indeed, all angiogenic markers investigated (vWf, CD31, VEGFa, Ve-cad, TGFβ1, CD44 and αSMA) were reduced in comparison to mice treated with control BM-MSCs.

Paclitaxel Uptake and Release by Different Stromal Cells

In the above reported experiments, MSCs expressing all the phenotypic and functional features that distinguish the MSCs phenotype from other stromal cells, have been used. However, MSCs represent only a subpopulation of the more complex stromal cells population in a tissue [41].

For this reason, in order to verify and support the idea that uptake and release of PTX may be considered a general character of stromal cells, we investigated whether other kinds of stromal cells primed with PTX behave like BM-MSCs. We studied both MSCs from human adipose tissue (AT-MSCs) and human skin derived fibroblasts (hSDFs). AT-MSCs confirmed the same differentiation potential (osteo, chondro and adipo) and the same markers as BM-MSCs (positivity for CD44, CD73, CD90, CD105, HLA-I and negativity for CD14, CD31, CD34, CD45, CD80 and HLA-II). hSDFs express a high level of typical MSCs markers (CD44, CD90 and CD105), are negative for hematopoietic markers CD31, CD34 and CD45 and express a minor differentiation capacity if compared to MSCs (only a few cells express osteoblasts markers but no cells are able to differentiate into adipocytes or chondrocytes).

AT-MSCs and hSDFs are sensitive to the proliferation block exerted by PTX (IC_{50} values were respectively 2.55 ± 1.02 ng/ml for AT-MSCs and 7.01 $\pm$ 2.17 ng/ml for hSDFs). However, the two cell types showed the same high resistant to the cytotoxic effect of the drug (IC_{50}> 10,000 ng/ml).

Table 1. Effect of BM-MSCs and BM-MSCsPTX on tumor growth *in vivo*

Experimental group	Tumor model	Ratio BM-MSCs/TCs	Tumor weight (grams±sd)	Tumor/ total#
Co-injection BM-MSCs and TCs				
Control	DU145	-	0.998 ± 0.276	8/8
+ PTX 2000 ng/ml		-	1.243 ± 0.311	5/5
+ BM-MSCs		1:5	1.043 ± 0.231	6/6
+BM-MSCsPTX		1:5	0.468 ± 0.102*	6/8
Control	MOLT-4	-	2.55 ± 0.430	5/5
+ PTX 2000 ng/ml		-	2.10 ± 0.212	3/3
+ BM-MSCs		1:5	2.32 ± 0.456	5/5
+BM-MSCsPTX		1:5	-**	0/5
Control	B16	-	1.016 ± 0.230	10/10
+ PTX 2000 ng/ml		-	1.105 ± 0.212	6/6
+ BM-MSCs		1:5	1.118 ± 0.356	10/10
+BM-MSCsPTX		1:5	0.657 ± 0.245*	9/9
Intra-tumor injection of BM-MSCs				
Control	MOLT-4	-	4.78±1.67	6/6
+ PTX 2000 ng/ml		-	3.12± 1.98	5/5
+ BM-MSCs		$3x10^5$ cellsx4 times	3.01±1.79	6/6
+BM-MSCsPTX		$3x10^5$ cellsx4 times	0.78±0.72**	5/6

Number of mice with tumor (at the time of sacrifice) on total mice number.
* $p<0.05$, ** $p<0.01$ versus control (animals treated with TCs only)

AT-MSCs and hSDFs demonstrated the same capacity as BM-MSCs to uptake and release PTX *in vitro* but with different efficiency; indeed, mesenchymal stromal cells from adipose tissue showed a lower PEC (0.39±0.43 pg/cell) while hSDFs were able to uptake the drug in a comparable way to BM-MSCs (their PEC value was 0.86 ± 0.2 pg/cell). Even the *in vitro* assays both with CM and PTX primed cells on many tumor cell lines (MOLT-4, DU145, neuroblastoma GI-LI-N and SH-SY-5Y(+), osteosarcoma SK-ES-1 for AT-MSCs; MOLT-4, DU145, glioblastoma U87 MG, neuroblastoma SH-SY5Y(+) and LAN-5 for hSDFs) confirmed the antitumoral activity both of AT-MSCs and hSDFs [21, 22].

Conclusion

These results, taken together, disclose new insights into functional properties of mammalian stromal cells and also strongly support the possible use of stromal cells as carriers for chemotherapeutic drugs. From a toxicological point of view, these results support the hypothesis that these cells may play a dual role in organs by controlling drug toxicity both by reducing or improving drug toxicity, depending on their capacity "to adsorb" and subsequently to release a drug in its original form. Data from cancer patients receiving very high doses of chemotherapy seem to support this hypothesis which may also explain organ specific toxicity by chemotherapeutic agents [42, 43]. However, the most interesting indication arising from our studies is that MSCs isolated from different tissue sources can uptake and subsequently slowly release PTX. This suggests that MSCs may become an important new tool for drug delivery by increasing efficacy and targeting selectivity in cancer therapy.

Acknowledgments

These studies were partially supported by AIRC (Associazione Italiana Ricerca Cancro) Project AIRC IG-9062 and Fondazione Banco del Monte di Lombardia. The authors disclose any commercial affiliations and declare no conflict of interest related to the study and data presented. We thank Ms Andrea Smith for checking the English of the manuscript.

References

[1] Momin, EN; Mohyeldin, A; Zaidi, HA; Velam G; Quiñones-Hinojosa, A. Mesenchymal stem cells: new approaches for the treatment of neurological diseases. *Current Stem Cell Research and Therapy*, 2010, 5, 326–344.

[2] da Silva Meirelles, L; Chagastelles, PC; Nardi, NB. Mesenchymal stem cells reside in virtually all post-natal organs and tissues. *Journal of Cell Science*, 2006, 119, 2204–2213.

[3] Fukuchi, Y; Nakajima, H; Sugiyama, D; Hirose, I; Kitamura, T; Tsuji, K. Human placenta-derived cells have mesenchymal stem/progenitor cell potential. *Stem Cells*, 2004, 22, 649-658.

[4] Romanov, YA; Svintsitskaya, VA; Smirnov, VN. Searching for alternative sources of postnatal human mesenchymal stem cells:

candidate MSC-like cells from umbilical cord. *Stem Cells*, 2003, 21, 105–110.

[5] Miura, M; Gronthos, S; Zhao, M; Lu, B; Fisher, LW; Robey, PG; Shi, S. SHED: stem cells from human exfoliated deciduous teeth. *Proceedings of the National Academy of Sciences of the United States of America*, 2003, 100, 5807-5812.

[6] Shah, K. Mesenchymal stem cells engineered for cancer therapy. *Advanced drug delivery reviews*, 2012, 64, 739-748.

[7] Kern, S; Eichler, H; Stoeve, J; Klüter, H; Bieback, K. Comparative analysis of mesenchymal stem cells from bone marrow, umbilical cord blood, or adipose tissue. *Stem Cells*, 2006, 24, 1294-1301.

[8] Blasi, A; Martino, C; Balducci, L; Saldarelli, M; Soleti, A; Navone, SE; Canzi, L; Cristini, S; Invernici, G; Parati, EA; Alessandri, G. Dermal fibroblasts display similar phenotypic and differentiation capacity to fat-derived mesenchymal stem cells, but differ in anti-inflammatory and angiogenic potential. *Vascular Cell*, 2011, 8, 3-5.

[9] Wagner, W; Wein, F; Seckinger, A; Frankhauser, M; Wirkner, U; Krause, U; Blake, J; Schwager, C; Eckstein, V; Ansorge, W; Ho, AD. Comparative characteristics of mesenchymal stem cells from human bone marrow, adipose tissue, and umbilical cord blood. *Experimental Hematology*, 2005, 33, 1402-1416.

[10] Weldon, JE; Xiang, L; Chertov, O; Margulies, I; Kreitman, RJ; FitzGerald, DJ; Pastan, I. A protease-resistant immunotoxin against CD22 with greatly increased activity against CLL and diminished animal toxicity. *Blood*, 2009, 113, 3792-3800.

[11] Dhar, S; Gu, FX; Langer, R; Farokhzad, OC; Lippard, SJ. Targeted delivery of cisplatin to prostate cancer cells by aptamer functionalized Pt (IV) prodrug-PLGA-PEG nanoparticles. *Proceedings of the National Academy of Sciences of the United States of America*, 2008, 105, 17356-17361.

[12] Loebinger, MR; Eddaoudi, A; Davies, D; Janes, SM. Mesenchymal stem cell delivery of TRAIL can eliminate metastatic cancer. *Cancer Research*, 2009, 69, 4134-4142.

[13] Zhang, XB; Beard, BC; Trobridge, GD; Wood, BL; Sale, GE; Sud, R; Humphries, RK; Kiem, HP. High incidence of leukemia in large animals after stem cell gene therapy with a HOXB4-expressing retroviral vector. *The Journal of Clinical Investigation*, 2008, 118, 1502-1510.

[14] Noble, M. Can neural stem cells be used as therapeutic vehicles in the treatment of brain tumors? *Nature Medicine*, 2000, 6, 369– 370.

[15] Cocke, KS; Updyke, LW; Wierda, D. Mouse bone marrow stromal cells can accumulate and release cytotoxic concentrations of doxorubicin into the hematopoietic microenvironment. *In Vitro Toxicology*, 1993, 6, 243-256.

[16] Pessina, A; Piccirillo, M; Mineo, E; Catalani, P; Gribaldo, L; Marafante, E; Neri, MG; Raimondi, A. Role of SR4987 stromal cells in the modulation of doxorubicin toxicity to in vitro granulocyte-macrophage progenitors (CFU-GM). *Life Sciences*, 1999, 65, 513-523.

[17] Pessina, A; Bonomi, A; Coccè, V; Invernici, G; Navone, S; Cavicchini, L; Sisto, F; Ferrari, M; Viganò, L; Locatelli, A; Ciusani, E; Cappelletti, G; Cartelli, D; Arnaldo, C; Parati, E; Marfia, G; Pallini, R; Falchetti, ML; Alessandri, G. Mesenchymal stromal cells primed with paclitaxel provide a new approach for cancer therapy. *PLoS One*, 2011, 6, e28321.

[18] Pessina, A; Coccè, V; Pascucci, L; Bonomi, A; Cavicchini, L; Sisto, F; Ferrari, M; Ciusani, E; Crovace, A; Falchetti, ML; Zicari, S; Caruso, A; Navone, S; Marfia, G; Benetti, A; Ceccarelli, P; Parati, E; Alessandri, G. Mesenchymal stromal cells primed with Paclitaxel attract and kill leukaemia cells, inhibit angiogenesis and improve survival of leukaemia-bearing mice. *British journal of haematology*, 2013, 160, 766-778.

[19] Schiff, PB; Fant, J; Horwitz, SB. Promotion of microtubule assembly in vitro by Taxol. *Nature*, 1979, 277, 665-667.

[20] Belotti, D; Vergani, V; Drudis, T; Borsotti, P; Pitelli, MR; Viale, G; Giavazzi, R; Taraboletti, G. The microtubule-affecting drug paclitaxel has antiangiogenic activity. *Clinical Cancer Research*, 1996, 2, 1843-1849.

[21] Bonomi, A; Coccè, V; Cavicchini, L; Sisto, F; Dossena, M; Balzarini, P; Portolani, N; Ciusani, E; Parati, E; Alessandri, G; Pessina, A. Adipose-tissue-derived stromal cells primed in vitro with paclitaxel acquires anti-tumor activity. *International journal of immunopathology and pharmacology*, 2013, in press.

[22] Pessina, A; Coccè, V; Bonomi, A; Cavicchini, L; Sisto, F; Ferrari, M; Ciusani, E; Navone, S; Marfia, G; Parati, E; Alessandri, G. Human Skin-Derived Fibroblasts Acquire In Vitro Anti-Tumor Potential After Priming with Paclitaxel. *Anticancer Agents in Medicinal Chemistry*, 2013, 3, 523-530.

[23] Lorenz, K; Sicker, M; Schmelzer, E; Rupf, T; Salvetter, J; Schulz-Siegmund, M; Bader, A. Multilineage differentiation potential of human

dermal skin-derived fibroblasts. *Experimental Dermatology*, 2008, 17, 925-932.

[24] Roni, V; Habeler, W; Parenti, A; Indraccolo, S; Gola, E; Tosello, V; Cortivo, R; Abatangelo, G; Chieco-Bianchi, L; Amadori, A. Recruitment of human umbilical vein endothelial cells and human primary fibroblasts into experimental tumors growing in SCID mice. *Experimental Cell Research*, 2003, 287, 28-38.

[25] Pittenger, MF; Mackay, AM; Beck, SC; Jaiswal, RK; Douglas, R; Mosca, JD; Moorman, MA; Simonetti, DW; Craig, S; Marshak, DR. Multilineage potential of adult human mesenchymal stem cells. *Science*, 1999, 284, 143-147.

[26] Li, J; Law, HKW; Lau, YL; Chan, CGF. Differential damage and recovery of human mesenchymal stem cells after exposure to chemotherapeutic agents, *British journal of haematology*, 2004, 127, 326-334.

[27] Polioudaki, H; Kasrinaki, MC; Papadaki, HA; Theodoropoulos, PA. Microtubule-interacting drugs induce moderate and reversible damage to human bone marrow mesenchymal stem cells, *Cell Proliferation*, 2009, 42, 434-447.

[28] Molinari, A; Calcabrini, A; Meschini, S; Stringaro, A; Del Bufalo, D; Cianfriglia, M; Arancia, G. Detection of P-glycoprotein in the Golgi apparatus of drug-untreated human melanoma cells. *International Journal of Cancer*, 1998, 75, 885-893.

[29] Zhang, P; Tao, DD; Feng, YD; Xie, DX; Zhou, JF; Gong, JP. Paclitaxel induces apoptosis of acute leukaemia cells in S phase. *Ai Zheng*, 2006, 25, 1243-1246.

[30] Singla, AK; Garg, A; Aggarwal, D. Paclitaxel and its formulations. *International Journal of Pharmaceutics*, 2002, 235, 179-192.

[31] Mueller, LP; Luetzkendorf, J; Mueller, T; Reichelt, K; Simon, H; Schmoll, HJ. Presence of mesenchymal stem cells in human bone marrow after exposure to chemotherapy: evidence of resistance to apoptosis induction. *Stem Cells*, 2006, 24, 2753-2765.

[32] Chaudhary, PM; Roninson, IB. Induction of multidrug resistance in human cells by transient exposure to different chemotherapeutic drugs. *Journal of the National Cancer Institute*, 1993, 85, 632-639.

[33] Kane, SE; Pastan, I; Gottesman, MM. Genetic basis of multidrug resistance of tumor cells. *Journal of Bioenergetics and Biomembranes*, 1990, 22, 593-618.

[34] Nobili, S; Landini, I; Giglioni, B; Mini, E. Pharmacological strategies for overcoming multidrug resistance. *Current Drug Targets*, 2006, 7, 861–879.

[35] Islam, MN; Das, SR; Emin, MT; Wei, M; Sun, L; Westphalen, K; Rowlands, DJ; Quadri, SK; Bhattacharya, S; Bhattacharya, J. Mitochondrial transfer from bone-marrow-derived stromal cells to pulmonary alveoli protects against acute lung injury. *Nature Medicine*, 2012, 18, 759-765.

[36] Prockop, DJ. Mitochondria to the rescue. *Nature Medicine*, 2012, 18, 653-654.

[37] Lai, RC; Yeo, RW; Tan, KH; Lim, SK. Exosomes for drug delivery - a novel application for the mesenchymal stem cell. *Biotechnology Advances*, 2012, in press.

[38] van Dommelen, SM; Vader, P; Lakhal, S; Kooijmans, SA; van Solinge, WW; Wood, MJ; Schiffelers, RM. Microvesicles and exosomes: opportunities for cell-derived membrane vesicles in drug delivery. *Journal of Controlled Release*, 2012, 16, 635-644.

[39] Kunstfeld, R; Wickenhauser, G; Michaelis, U; Teifel, M; Umek, W; Naujoks, K; Wolff, K; Petzelbauer, P. Paclitaxel encapsulated in cationic liposomes diminishes tumor angiogenesis and melanoma growth in a "humanized" SCID mouse model. *The Journal of Investigative Dermatology*, 2003, 120, 476-482.

[40] Nicosia, RF; Ottinetti, A. Growth of microvessel in serum-free matrix culture of rat aorta: a quantitative assay of angiogenesis in vitro. *Laboratory Investigations*, 1990, 63, 115-122.

[41] Tormin, A; Brune, JC; Olsson, E; Valcich, J; Neuman, U; Olofsson, T; Jacobsen, SE; Scheding S. Characterization of bone marrow-derived mesenchymal stromal cells (MSC) based on gene expression profiling of functionally defined MSC subsets. *Cytotherapy*, 2009, 11, 114-128.

[42] Breeden, JH; Vollmer, JT; Twomey, PL. Toxicity of very high dose nitrosourea administration. *Cancer*, 1982, 50, 1728-1733.

[43] Gille, L; Nohl, H. Analyses of the molecular mechanism of adriamycin-induced cardiotoxicity. *Free Radical Biology and Medicine*, 1997, 23, 775-782.

In: Paclitaxel
Editor: Diego Morales
ISBN: 978-1-62808-549-5

Chapter 6

The Role of Paclitaxel in Ovarian Cancer

Mariko Shindo, Kiyoshi Yoshino*, Yutaka Ueda and Tadashi Kimura

Department of Obstetrics and Gynecology, Osaka University, Graduate School of Medicine, Yamadaoka, Suita, Osaka Japan

Abstract

Epithelial ovarian carcinoma, although rare, causes more deaths than any other female genital tract cancer and is overall the fifth most frequent cause of cancer death in women. Its high lethality is because more than two-thirds of patients have a highly advanced disease at diagnosis. In the United States, there were approximately 22,280 new cases and 15,500 deaths in 2012 as a result of ovarian cancer [1]. Systemic chemotherapy plays an important role in the treatment of epithelial ovarian cancer. After the introduction of cisplatin in the latter half of the 1970s, platinum-based combination chemotherapy became the regimen of choice. The incorporation of paclitaxel into chemotherapeutic regimens in late 1990s was a major advance in the treatment of advanced stage disease. A series of randomized prospective clinical trials with paclitaxel-containing arms

* Corresponding author: Dr. Kiyoshi Yoshino, M.D., Ph.D. Department of Obstetrics and Gynecology, Osaka University, Graduate School of Medicine, 2-2, Yamadaoka, Suita, Osaka 565-0871Japan, Tel: +81-66879-3355, Fax: +81-66879-3359, E-mail: yoshino@gyne.med.osaka-u.ac.jp.

defined the combination of intravenous carboplatin and paclitaxel as the standard for first-line chemotherapy for ovarian cancer, to the point where now ovarian cancer treatment cannot be started without paclitaxel.

Surgery

The primary treatment for ovarian cancer consists of an appropriate surgical staging and cytoreduction, followed in most patients by systemic chemotherapy. The operation phase typically includes a total abdominal hysterectomy and bilateral salpingo-oophorectomy, along with a complete omentectomy, retroperitoneal lymphadnectomy and resection of any visible metastatic lesions [2-5]. The amount of residual tumor after this surgery is one of the most important predictors of survival for woman with epithelial ovarian cancer [6]. Surgical cytoreduction is classified as optimal if the residual tumor nodules are all less than 1 cm in maximal diameter [7-11]. Procedures that may be considered for optimal surgery include: radical pelvic dissection, bowel resection, diaphragm or other peritoneal surface stripping, splenectomy, partial hepatectomy, etc.

Paclitaxel in First-Line Chemotherapy

In treatment for epithelial ovarian cancer, most patients receive postoperative systemic chemotherapy. Because, with surgical treatment alone, survival is greater than 90% for patients with early stage, low risk tumors (stage IA or IB, grade1 tumors), observation only is recommended for this group [12]. In patients with more advanced-stage disease, systemic multi-agent chemotherapy is the standard.

Historically, ovarian cancer was one of the first malignant solid tumors to demonstrate responsiveness to chemotherapy. After the introduction of cisplatin in the 1970s, platinum-based combination chemotherapy became the most frequently used treatment regimen for this tumor, and both response rates and median survivals were significantly increased.

The next major advance in the treatment of advanced-stage disease was the incorporation of mitotic inhibitor paclitaxel into chemotherapeutic regimens in the late 1990s. Following the observations that paclitaxel produced positive objective responses in16% to 37% of women with relapsed ovarian cancer [13-16], platinum-paclitaxel combination regimens were developed.

A series of randomized studies in advanced ovarian cancer were completed, in which a paclitaxel-containing arm was compared to a non-paclitaxel arm. In the GOG 111 trial, 386 women with sub-optimally debulked stage III or stage IV disease were randomly assigned to receive six cycles of cisplatin (75 mg/m^2) plus paclitaxel (135 mg/m^{2}over 24 h) or cisplatin (75 mg/m^2) plus cyclophosphamide (750 mg/m^2)[17].The paclitaxel-containing regimen was associated with significant improvements in overall response (73%vs. 60%), clinical complete responses (CR; 51%vs. 31%), median progression-free survival (PFS; 18 vs. 13 months) and overall survival (OS; 38 vs. 24 months). After more than 5 years of follow-up, paclitaxel treatment was still associated with a 28% reduction in the risk of progression and a 34% reduction in the risk of death.

The European–Canadian cooperation within the OV-10 trial confirmed these results; 680 patients were randomly treated with cisplatin (75 mg/m^2) and paclitaxel (175 mg/m^{2}over 3 h), or with cisplatin (75 mg/m^2) plus cyclophosphamide (750 mg/m^2) [18]. In total, 48% of the women in the cisplatin–cyclophosphamide arm in this intergroup trial were crossed over to paclitaxel at progression of disease. The paclitaxel–platinum arm was associated with a significant improvement in overall response (59%vs. 45%), clinical CR (41%vs. 27%) compared with cisplatin–cyclophosphamide. After a median follow-up of 39 months, paclitaxel/cisplatin also showed a significant advantage in median PFS (16 vs. 12 months) and OS (36 vs. 26 months), despite the high rate of crossover. Based on these two studies, it was concluded that paclitaxel should be included in the primary treatment of all women with advanced-stage epithelial ovarian cancer and the cisplatin–paclitaxel combination was considered to be the worldwide standard for first-line ovarian cancer chemotherapy.

However, in view of cisplatin mediated gastrointestinal and neurologic toxicity, as well as the need for hydration to avoid renal toxicity, and the inconvenient 24-h infusion schedule for paclitaxel, there was considerable interest in switching to a different regimen, one with carboplatin and short-infusion of paclitaxel [19, 20]. Thus, GOG 158 was opened in 1995 and closed in 1998 with the primary objective of providing non-inferiority evidence for carboplatin/paclitaxel compared with cisplatin/paclitaxel in optimally resected (<1 cm residual disease) stage III ovarian carcinoma. Seven hundred ninety-two eligible patients were randomized to cisplatin (75 mg/m^2) and a 24-hour infusion of paclitaxel (135 mg/m^2) intravenously (IV) or carboplatin AUC 7.5

and 3-hour infusion of paclitaxel (175mg/m^2) IV every 3 weeks for 6 cycles [21].

Although the median PFS and OS were not statistically significantly different between the 2 groups, there was a survival advantage to carboplatin/paclitaxel (median PFS 19 vs. 21 months; OS49vs. 57 months), for cisplatin/paclitaxel vs. carboplatin/paclitaxel, respectively. The relative risk for progression and death were 0.88 (95% CI, 0.75-1.03) and 0.84 (95% CI, 0.70-1.02) for the carboplatin/paclitaxel group. In addition, grade 4 leukopenia, gastrointestinal, renal, and metabolic toxicity were higher in the cisplatin /paclitaxel arm.

Therefore, the investigators concluded that the carboplatin/paclitaxel combination was not inferior to cisplatin/paclitaxel and resulted in less toxicity than cisplatin/paclitaxel. Similar results was obtained in a large randomized trial in Germany, in which the dose of carboplatin was AUC=6 and paclitaxel was 185mg/m^2 over 3 hours, compared with the same dose of paclitaxel and cisplatin 75 mg/m^2 [22]. Based on these data, the preferred regimen for patients with advanced-stage disease is the paclitaxel plus carboplatin combination.

Paclitaxel in Dose-Dense Chemotherapy

The concept of dose density is based on the hypothesis that a shorter interval between doses of cytotoxic therapy will more effectively reduce tumor burden than will dose escalation. Although dose intensification of platinum in ovarian cancer has been studied in the past, there was no improvement in prognosis shown.

Katsumata and colleagues, from Japanese Gynecologic Oncology Group, reported a phase III study comparing the conventional 3-weekly paclitaxel and carboplatin schedule versus dose-dense weekly paclitaxel and 3-weekly carboplatin. Participants were randomized to receive either the dose-dense regimen of carboplatin (AUC 6.0 on day 1 and paclitaxel 80 mg/m^{2}on days 1, 8, and 15) versus the standard regimen of carboplatin (AUC 6.0 and paclitaxel 180 mg/m^{2}on day 1) [23]. Treatments were repeated every 21 days for six cycles. Patients who responded well were eligible for three additional cycles. After 6.4 years of follow-up, the 312 patients who received dose-dense therapy achieved a median progression-free survival (PFS) of 28.2 months, versus a median PFS of only 17.5 months for the 319 patients who received the standard regimen (HR=0.76, 95% CI [0.62-0.91]; p=0.0037). The median overall survival (OS) had not yet been reached in the dose-dense group, while the OS among those treated conventionally was 62.2 months. At five years, the survival rate was 58.7% in the dose-dense group, compared with 51.1% in the standard-treatment group (HR=0.79, 95% CI, 0.63-0.99; p=0.039). While the study provides insight into dosing strategies, questions remain unanswered regarding the impact of the route of administration. Ongoing studies will compare the effect of weekly paclitaxel on the overall survival benefit with that of using intraperitoneal chemotherapy.

Paclitaxel Combined with Intraperitoneal (IP) Chemotherapy

The most common route of ovarian cancer spread is within the peritoneal cavity. The benefit of administering chemotherapy directly into the peritoneal cavity is supported by preclinical and pharmacokinetic data. Compared to intravenous (IV) treatment, intraperitoneal (IP) administration permits a several-fold increase in drug concentration to be achieved within the abdominal cavity [24].The intraperitoneal route of administration for cisplatin and intravenous paclitaxel chemotherapy in the primary treatment of optimally resected stage III ovarian cancer is recommended, based on randomized control trials. In a randomized prospective GOG study(protocol 172), the standard arm, consisting of a combination of cisplatin (75mg/m^2)and paclitaxel (135mg/m^2) given intravenously every 3 weeks, was compared with an experimental arm consisting of paclitaxel (135 mg/m^2) (IV)on day1, followed by cisplatin (100mg/m^2)(IP) on day 2, and paclitaxel (60mg/m^2)(IP) on day 8, every 3 weeks, each given for six cycles [25].

In this study, 429 patients were randomly assigned and 415 were eligible. The median progression-free survival was 23.8 months in the IP arm, versus 18.3 months in the IV arm (p=0.05). The median overall survival was 65.6 months in the IP arm, and 49.7 months in the IV arm (p=0.03). The complication rate of the experimental IP regimen was significantly greater than the complication rate of IV therapy alone. The IP chemotherapy group had more severe toxicities, which included pain and fatigue and hematologic, gastrointestinal, metabolic, and neurologic toxicities. Quality of life was lower in the IP chemotherapy group during the first year after initial treatment. In addition, only 42% of the patients randomized to IP chemotherapy completed the recommended six cycles. The most common reason for discontinuation of IP therapy was catheter–related problems.

Based on this trial, together with the previous studies, the combination of paclitaxel (IV) plus platinum (IP) chemotherapy may be an option for well counseled, carefully selected patients with optimally debulked stage III ovarian cancer, when provided by a physician with the requisite training and experience in administering this treatment. However, given the balance of efficacy, quality of life, and toxicity, the decision to use IP chemotherapy must still be individualized [26, 27].

Paclitaxel in Recurrent disease

Despite initial therapy (usually consisting of surgical cytoreduction and platinum-taxane combination therapy), the majority of women with advanced stage ovarian cancer will relapse and require additional treatment. Taking into account the frequency of each stage of disease and its projected relapse rate, the overall likelihood of relapse after initial therapy for all stages of disease for women with EOC is 62%; it is 80 to 85% for women who present with stage III or IV disease. The management of relapsed disease is frequently stratified based upon the amount of time that has elapsed between the completion of platinum-based treatment and the detection of the relapse, known as the platinum-free interval (PFI) [28-30]. Women who relapse longer than 6 months after completion of primary chemotherapy are classified as platinum-sensitive and some studies suggest that a combination of platinum plus paclitaxel may be better in these patients for second-line therapy than platinum-alone. In parallel phase III trials conducted by the International Collaboration on Ovarian Neoplasms (ICON-4) and the Arbeitsgemeinschaft Gynaekologische Onkologie Studiengruppe Ovarialkarzinom (AGO-OVAR-2.2) [31, 32], 802 women were treated with either a single platinum agent (carboplatin or cisplatin) or a non-taxane containing platinum combination (e.g. cisplatin, doxorubicin and cyclophosphamide, CAP), or paclitaxel plus a platinum agent (with 80 percent of cases receiving carboplatin).

Compared to use of a non-paclitaxel containing therapy, treatment with paclitaxel plus a platinum agent (with a median follow-up of 42 months) resulted in a significant survival benefit (hazard ratio, 0.82, 95% CI, 0.69-0.97),a five month improvement in median survival (29 vs. 24 months). The toxicities were comparable, except there was a significantly higher incidence of neurologic toxicity and alopecia in the paclitaxel group, whereas myelosuppression was significantly greater with the non-paclitaxel-containing regimens.

These data support the benefit of combination therapy; however, single agent therapy may be appropriate for medically frail patients, those with lingering toxicities from prior therapy, or a history of a hypersensitivity reaction. In platinum-refractory patients, who relapse less than 6 months after completion of initial chemotherapy, response rates to second-line chemotherapy are less than 10% and the median progression-free survival is short, around 4 months.

Single-agent therapy (i.e., paclitaxel, docetaxel, oral etoposide, gemcitabine, liposomal doxorubicin, topotecan) is typically used because combination regimens are associated with more toxicity without any apparent additional benefit [33-39]. Single-agent paclitaxel shows objective responses in 20-30% in phase II trials of women with platinum-resistant ovarian cancer [13-16].

The main toxicities are fatigue and peripheral neuropathy. Weekly administration of paclitaxel is active, and the toxicity, especially myelosuppression, is mild. In a study of 53 women with platinum-resistant ovarian cancer with measurable disease, weekly paclitaxel (80mg/m^2) had an objective response of 25% [40].

Nab-paclitaxel is a novel clinical entity incorporating paclitaxel into an albumin nanoparticle. In phase II trials of 47 women with platinum-resistant ovarian cancer, receiving nab-paclitaxel (100mg/m^2,on days 1, 8, and 15, on a 28-day schedule), the response rate was 23%. Median progression-free survival was 4.5 months and overall survival was 17.4 months. This study supported the concept that weekly treatment with nab-paclitaxel is associated with clinical efficacy and a tolerable side effect [41, 42].

Conclusion

In the past two decades, paclitaxel have emerged as a front-line agent for ovarian cancer chemotherapy, along with the platinum agents. Having already learned about the efficacy of paclitaxel in ovarian cancer, finding the optimal dosing schedule, identifying the optimal route of administration, and elucidating the mechanisms of taxane resistance are areas of intense research. Future trials will continue to define the optimal role of paclitaxel in the treatment of ovarian cancer.

References

[1] Siegel, R., D. Naishadham, and A. Jemal, Cancer statistics. CA *Cancer J. Clin*, 2012. 62(1): p. 10-29.

[2] Bristow, R.E., et al., Survival effect of maximal cytoreductive surgery for advanced ovarian carcinoma during the platinum era: a meta-analysis. *J. Clin. Oncol.*, 2002. 20(5): p. 1248-59.

[3] du Bois, A., et al., 2004 consensus statements on the management of ovarian cancer: final document of the 3rd International Gynecologic Cancer Intergroup Ovarian Cancer Consensus Conference (GCIG OCCC 2004). *Ann. Oncol.*, 2005. 16 Suppl 8: p. viii7-viii12.

[4] du Bois, A., et al., Role of surgical outcome as prognostic factor in advanced epithelial ovarian cancer: a combined exploratory analysis of 3 prospectively randomized phase 3 multicenter trials: by the Arbeitsgemeinschaft Gynaekologische Onkologie Studiengruppe Ovarialkarzinom (AGO-OVAR) and the Groupe d'Investigateurs Nationaux Pour les Etudes des Cancers de l'Ovaire (GINECO). *Cancer,* 2009. 115(6): p. 1234-44.

[5] Eisenhauer, E.L., et al., The addition of extensive upper abdominal surgery to achieve optimal cytoreduction improves survival in patients with stages IIIC-IV epithelial ovarian cancer. *Gynecol. Oncol.*, 2006. 103(3): p. 1083-90.

[6] du Bois, A., et al., Potential role of lymphadenectomy in advanced ovarian cancer: a combined exploratory analysis of three prospectively randomized phase III multicenter trials. *J. Clin. Oncol*, 2010. 28(10): p. 1733-9.

[7] Aletti, G.D., et al., Aggressive surgical effort and improved survival in advanced-stage ovarian cancer. *Obstet. Gynecol.,* 2006. 107(1): p. 77-85.

[8] Bristow, R.E., I. Puri, and D.S. Chi, Cytoreductive surgery for recurrent ovarian cancer: a meta-analysis. *Gynecol. Oncol.*, 2009. 112(1): p. 265-74.

[9] Eisenhauer, E.L., et al., The effect of maximal surgical cytoreduction on sensitivity to platinum-taxane chemotherapy and subsequent survival in patients with advanced ovarian cancer. *Gynecol. Oncol.*, 2008. 108(2): p. 276-81.

[10] Elattar, A., et al., Optimal primary surgical treatment for advanced epithelial ovarian cancer. *Cochrane Database Syst. Rev.*2011, (8): p. CD007565.

[11] Fader, A.N. and P.G. Rose, Role of surgery in ovarian carcinoma. *J. Clin. Oncol.*, 2007. 25(20): p. 2873-83.
[12] Young, R.C., et al., Adjuvant therapy in stage I and stage II epithelial ovarian cancer. Results of two prospective randomized trials. *N. Engl. J. Med.*, 1990. 322(15): p. 1021-7.
[13] Ghamande, S., et al., Weekly paclitaxel in patients with recurrent or persistent advanced ovarian cancer. *Int. J. Gynecol. Cancer,* 2003. 13(2): p. 142-7.
[14] Kohn, E.C., et al., Dose-intense taxol: high response rate in patients with platinum-resistant recurrent ovarian cancer. *J. Natl. Cancer Inst.*, 1994. 86(1): p. 18-24.
[15] Markman, M., et al., Phase II trial of weekly single-agent paclitaxel in platinum/paclitaxel-refractory ovarian cancer. *J. Clin. Oncol.*, 2002. 20(9): p. 2365-9.
[16] Omura, G.A., et al., Phase III trial of paclitaxel at two dose levels, the higher dose accompanied by filgrastim at two dose levels in platinum-pretreated epithelial ovarian cancer: an intergroup study. *J. Clin. Oncol.*, 2003. 21(15): p. 2843-8.
[17] McGuire, W.P., et al., Cyclophosphamide and cisplatin versus paclitaxel and cisplatin: a phase III randomized trial in patients with suboptimal stage III/IV ovarian cancer (from the Gynecologic Oncology Group). *Semin. Oncol.,* 1996. 23(5 Suppl 12): p. 40-7.
[18] Piccart, M.J., et al., Randomized intergroup trial of cisplatin-paclitaxel versus cisplatin-cyclophosphamide in women with advanced epithelial ovarian cancer: three-year results. *J. Natl. Cancer Inst.*, 2000. 92(9): p. 699-708.
[19] Bookman, M.A., et al., Carboplatin and paclitaxel in ovarian carcinoma: a phase I study of the Gynecologic Oncology Group. *J. Clin. Oncol.*, 1996. 14(6): p. 1895-902.
[20] Eisenhauer, E.A., et al., European-Canadian randomized trial of paclitaxel in relapsed ovarian cancer: high-dose versus low-dose and long versus short infusion. *J. Clin. Oncol.,* 1994. 12(12): p. 2654-66.
[21] Ozols, R.F., et al., Phase III trial of carboplatin and paclitaxel compared with cisplatin and paclitaxel in patients with optimally resected stage III ovarian cancer: a Gynecologic Oncology Group study. *J. Clin. Oncol.*, 2003. 21(17): p. 3194-200.
[22] du Bois, A., et al., A randomized clinical trial of cisplatin/paclitaxel versus carboplatin/paclitaxel as first-line treatment of ovarian cancer. *J. Natl. Cancer Inst.,* 2003. 95(17): p. 1320-9.

[23] Katsumata, N., et al., Dose-dense paclitaxel once a week in combination with carboplatin every 3 weeks for advanced ovarian cancer: a phase 3, open-label, randomised controlled trial. *Lancet,* 2009. 374(9698): p. 1331-8.
[24] Markman, M., Intraperitoneal chemotherapy. *Semin. Oncol.*, 1991. 18(3): p. 248-54.
[25] Armstrong, D.K., et al., Intraperitoneal cisplatin and paclitaxel in ovarian cancer. *N. Engl. J. Med.*, 2006. 354(1): p. 34-43.
[26] Jaaback, K. and N. Johnson, Intraperitoneal chemotherapy for the initial management of primary epithelial ovarian cancer. *Cochrane Database Syst. Rev.*, 2006(1): p. CD005340.
[27] Walker, J.L., et al., Intraperitoneal catheter outcomes in a phase III trial of intravenous versus intraperitoneal chemotherapy in optimal stage III ovarian and primary peritoneal cancer: a Gynecologic Oncology Group Study. *Gynecol. Oncol.*, 2006. 100(1): p. 27-32.
[28] Gore, M.E., et al., Treatment of relapsed carcinoma of the ovary with cisplatin or carboplatin following initial treatment with these compounds. *Gynecol. Oncol.,* 1990. 36(2): p. 207-11.
[29] Markman, M., Second-line therapy for potentially platinum-sensitive recurrent ovarian cancer: what is optimal treatment? *Gynecol. Oncol.,* 2001. 81(1): p. 1-2.
[30] Markman, M., et al., Duration of response to second-line, platinum-based chemotherapy for ovarian cancer: implications for patient management and clinical trial design. *J. Clin. Oncol.,* 2004. 22(15): p. 3120-5.
[31] Gonzalez-Martin, A.J., et al., Randomized phase II trial of carboplatin versus paclitaxel and carboplatin in platinum-sensitive recurrent advanced ovarian carcinoma: a GEICO (Grupo Espanol de Investigacion en Cancer de Ovario) study. *Ann. Oncol.*, 2005. 16(5): p. 749-55.
[32] Parmar, M.K., et al., Paclitaxel plus platinum-based chemotherapy versus conventional platinum-based chemotherapy in women with relapsed ovarian cancer: the ICON4/AGO-OVAR-2.2 trial. *Lancet,* 2003. 361(9375): p. 2099-106.
[33] Ferrandina, G., et al., Phase III trial of gemcitabine compared with pegylated liposomal doxorubicin in progressive or recurrent ovarian cancer. *J. Clin. Oncol.*, 2008. 26(6): p. 890-6.
[34] Gordon, A.N., et al., Long-term survival advantage for women treated with pegylated liposomal doxorubicin compared with topotecan in a

phase 3 randomized study of recurrent and refractory epithelial ovarian cancer. *Gynecol. Oncol.*, 2004. 95(1): p. 1-8.

[35] Miller, D.S., et al., Phase II evaluation of pemetrexed in the treatment of recurrent or persistent platinum-resistant ovarian or primary peritoneal carcinoma: a study of the Gynecologic Oncology Group. *J. Clin. Oncol.*, 2009. 27(16): p. 2686-91.

[36] Mutch, D.G., et al., Randomized phase III trial of gemcitabine compared with pegylated liposomal doxorubicin in patients with platinum-resistant ovarian cancer. *J. Clin. Oncol.*, 2007. 25(19): p. 2811-8.

[37] Rose, P.G., et al., A phase II study of docetaxel in paclitaxel-resistant ovarian and peritoneal carcinoma: a Gynecologic Oncology Group study. *Gynecol. Oncol.*, 2003. 88(2): p. 130-5.

[38] Rose, P.G., et al., Prolonged oral etoposide as second-line therapy for platinum-resistant and platinum-sensitive ovarian carcinoma: a Gynecologic Oncology Group study. *J. Clin. Oncol.*, 1998. 16(2): p. 405-10.

[39] Rothenberg, M.L., et al., Phase II trial of vinorelbine for relapsed ovarian cancer: a Southwest Oncology Group study. *Gynecol. Oncol.*, 2004. 95(3): p. 506-12.

[40] Greco, F.A. and J.D. Hainsworth, One-hour paclitaxel infusion schedules: a phase I/II comparative trial. *Semin. Oncol.*, 1995. 22(3 Suppl 6): p. 118-23.

[41] Coleman, R.L., et al., A phase II evaluation of nanoparticle, albumin-bound (nab) paclitaxel in the treatment of recurrent or persistent platinum-resistant ovarian, fallopian tube, or primary peritoneal cancer: a Gynecologic Oncology Group study. *Gynecol. Oncol, 2011.*122(1): p. 111-5.

[42] Teneriello, M.G., et al., Phase II evaluation of nanoparticle albumin-bound paclitaxel in platinum-sensitive patients with recurrent ovarian, peritoneal, or fallopian tube cancer. *J. Clin. Oncol.*, 2009. 27(9): p. 1426-31.

Index

A

F

G

H

M

N

O

P

Q

R

S